WORDS

OF THE

LAGOON

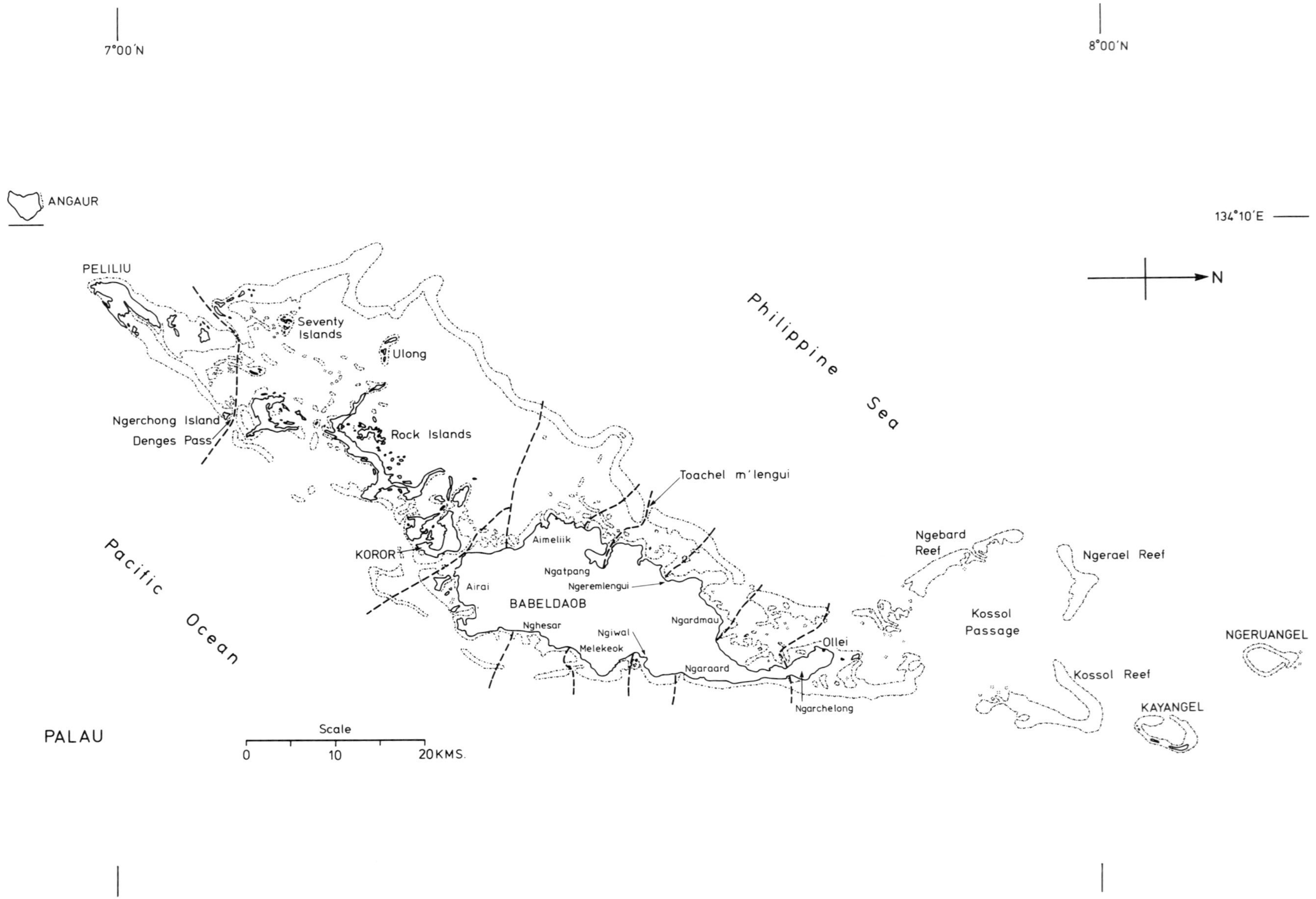

134°10'E

8°00'N

7°00'N

N

Philippine Sea

NGERUANGEL

Ngerael Reef

KAYANGEL

Ngebard Reef

Kossol Reef

Kossol Passage

Ollei

Ngarchelong

Toachel m'lengui

Ngardmau

Ngaraard

BABELDAOB

Ngatpang

Ngeremlengui

Aimeliik

Ngiwal

Melekeok

Airai

Nghesar

Ulong

Rock Islands

KOROR

Seventy Islands

Ngerchong Island

Denges Pass

PELILIU

ANGAUR

Pacific Ocean

PALAU

Scale

0   10   20KMS.

WORDS

OF THE

LAGOON

FISHING AND MARINE LORE IN THE PALAU DISTRICT OF MICRONESIA

R. E. JOHANNES

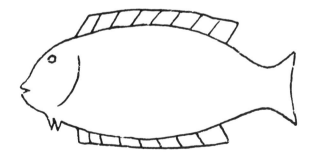

UNIVERSITY OF CALIFORNIA PRESS
BERKELEY • LOS ANGELES • LONDON

University of California Press
Berkeley and Los Angeles, California

University of California Press, Ltd.
London, England

First Paperback Printing 1992

Library of Congress Cataloging in Publication Data
Johannes, Robert Earl, 1936–
  Words of the lagoon.

  Contains glossaries of Palauan and Tobian words.
  Bibliography: p. 207
  Includes index.
  1. Fisheries—Palau Islands.   2. Marine biology—
Palau Islands.   3. Fishes—Palau Islands.   4. Folk-
lore—Palau Islands.   5. Ethnology—Palau Islands.
6. Palauan language—Terms and phrases.   7. Palau
Islands.   I. Title.
SH319.P44J64   639'.22'09966   79-64483
ISBN 0-520-08087-4

Designed by Graphics Two, Los Angeles
Printed in the United States of America

2 3 4 5 6 7 8 9

The paper used in this publication meets the minimun
requirements of American National Standard for Information Sciences—Permanence of Paper for Printed Library
Materials, ANSI Z39.48–1984. ♾

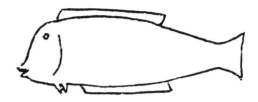

CONTENTS

PREFACE

Few people would claim to know as much about how to catch fish as a good full-time fisherman. When it comes to understanding fish behavior and the many environmental factors that help determine and predict it, marine biologists must often take a back seat. This is hardly surprising. There are hundreds of times as many fishermen today as there are marine biologists, and their forebears were plying their trade and passing on their accumulated knowledge tens of centuries before anyone heard of marine biology. What *is* surprising is how little effort has been made by scientists to search out and record this information.

Traditional native fishermen are especially rich sources of unrecorded knowledge. A modern commercial fisherman fifteen feet off the water in a rumbling trawler searches for his fish with machines. Isolated in his wheelhouse he perceives his prey as abstract shadows on an echo-sounder chart. The native fisherman searches with his eyes and ears. In shallow water he stalks fish at close range on foot. He pursues them in their own realm with goggles and spear. He knows the local currents intimately, for in his small, often motorless craft he must harness them when they are benign and avoid them when they are not. He is, in short, more in touch with his prey and their surroundings than his modern, mechanized counterpart. As Ommaney (1966) states, he "has forgotten more about how to catch the fishes of his particular bay or lagoon than we shall ever learn."

Native fishermen are particularly knowledgeable on small islands where seafood is the main source of animal protein and fishing is often the single most important male occupation. The waters around such islands, moreover, are generally influenced little by nutrient- and sediment-bearing terrestrial runoff. As a consequence they are typically very clear, affording fishermen with especially favorable conditions for observing the behavior of their prey. A particularly large reservoir of information about the sea and its inhabitants exists in the islands of the tropical Pacific because of their great number and the high biological diversity of their marine communities.

Pacific islanders' knowledge of fish behavior is, according to Ottino and Plessis (1972, p. 370), "of a stupefying richness, and at times of such precision that the corresponding poverty of our own conceptions makes inquiry very difficult." Gosline and Brock (1960, p. 1) state, "It is probable that the Hawaiians of Captain James Cook's time knew more about the fishes of their islands than is known today." Groves (1933–34, p. 432) said that, "as a result of his regular association with fish getting activities, the average male native [of Tabar Island, Bismarck Archipelago] has an amazing knowledge of the habits, type, and value for food purposes of the innumerable varieties of fish in the adjacent waters." Concerning Tahitians, Handy (1932, p. 77) wrote, "the native fisherman is possessed of a store of precise knowledge that may be truly characterized as a natural science," and Nordhoff (1930, p. 233) stated, "an accomplished fly fisherman in Europe or America does not carry in his head one-half the store of practical knowledge a bonito fisherman uses every day."

But despite such enthusiastic endorsements, very little serious effort has been made to collect and record this knowledge. Some anthropologists are well trained in biology and could carry out such research. But anthropologists interested in the ethnobiology of Oceania have focused largely on terrestrial ecosystems. This surprises me as a higher percentage of tradition is preserved in the Pacific islander's relations with his marine ecosystem than with his land (see, e.g., Danielsson, 1956). The terrestrial ecology of virtually every island in Oceania has been drastically altered by the introduction of foreign plants and animals and the extinction of indigenous ones (e.g., Fosberg, 1972). Marine communities have not undergone changes of comparable magnitude; there have been far fewer marine introductions and very few known marine extinctions.

To be sure, many pages have been written on Pacific island canoes and fishing implements and how they have been used to catch fish. But little has been written on why the islanders fish the way they do and what they know about their prey and its environment. This is probably because it is the nature of anthropology to focus mainly on people. Accordingly, when anthropologists study man-in-nature, the general form their queries take is usually, "how does this environment influence you?" rather than "what can we learn about this environment from you?" (There are some exceptions; e.g., Blurton Jones and Konner, 1976; Nelson, 1969).

The lack of interest shown by biologists in such knowledge has a different root. Natural scientists have routinely overlooked the practical knowledge possessed by artisans (e.g., Hanlon, 1979; Isaacs, 1976). It is one manifestation of the elitism and ethnocentrism that run deep in much of the Western scientific community. If unpublished notebooks containing the detailed observations of a long line of biologists and oceanographers were destroyed, we would be outraged. But when specialized knowledge won from the sea over centuries by formally unschooled but uniquely qualified observers—fishermen—is allowed to disappear as the westernization of their cultures proceeds, hardly anyone seems to care.

There seems to have been only one good published general study of Pacific island marine lore. It was written neither by a biologist nor an anthropologist, but by a writer, Charles Nordhoff, coauthor of *Mutiny on the Bounty*. Nordhoff fished extensively with Tahitians using their techniques and was a fine observer. His 1933 study of Tahitian offshore fishing has been called "a minor masterpiece stylistically and ethnographically," (Oliver, 1974, p. 284).

But offshore fishing involves only a few species of fish. Shallow coral reef and lagoon habitats, in contrast, yield several hundred species of edible marine animals, and this is where the expert native fisherman's knowledge, as Nordhoff acknowledged, is most awesome. "The time is ripe," he stated, "for some trained enthusiast to settle in these islands, learn the language, and devote four or five years to a complete account of fishing, inside and outside the reefs. Such a work would assume proportions almost encyclopaedic, and bring to light a mass of curious data."

The present work grew out of what is probably the first reasonably comprehensive attempt to carry out the kind of studies that Nordhoff called for forty years ago—to discover what Westerners can learn about tropical marine ecosystems and their resources by investigating the knowledge and actions of native fishermen and by observing their impact on these resources.

There are practical reasons for such studies, particularly in the tropics. Coral reef communities cover an estimated 230,000 square miles of shallow tropical sea bottom and appear to have a fin-fisheries potential of 6–7 million tons per year (Smith, 1978). This would be enough fish to feed the entire population of the United States at its current rate of fish consumption for about four years, and is equivalent to about 9 percent of the world's current commercial catch from the sea.

But this potential is not being realized. Many reef and lagoon fisheries appear to be heavily overexploited today, whereas many others are clearly underexploited. Scientific knowledge concerning these resources and their effective exploitation is meager. Moreover they present biologists with the most complex fisheries management problems in the world. There are far more species than in higher latitudes and the fisheries are not dominated by a few overwhelmingly important species as is the case in most temperate waters. Learning enough about the biology of the hundreds of species that contribute to the catch so as to manage their exploitation efficiently is an enormous undertaking. Even if tropical countries had the money, facilities, and trained scientists to carry on marine research programs of comparable strength to those in the richer countries of the temperate zone (an almost impossible dream for most of them in this century), it would take decades to amass enough information to manage the many important tropical marine species as efficiently as some shrimp, salmon, or halibut populations are managed in temperate waters today.

The kind of research described in this book offers a shortcut to some of the basic natural history data we need in order to understand these vast and valuable resources. Such information has to be quantified and blended with more sophisticated forms of biological research (e.g., population dynamics, behavior, physiology, genetics) before it can be put to optimum use, and this is no small matter. But I gained more new (to marine science) information during sixteen months of fieldwork using this approach than I had during the previous fifteen years using more conventional research techniques. This is because of my access to a store of unrecorded knowledge gathered by highly motivated observers over a period of centuries. This book, then, is really the work of uncounted individuals carried out over many generations.

For several years prior to this work I had been trying to deduce what sets the upper limits on the yield of fish to man in coral reef communities. Although coral reefs are among the most biologically productive communities on earth, their fish populations seem surprisingly vulnerable to overharvesting. I came to Palau with an ecological hypothesis to explain this. But after a few weeks I became aware of various political, cultural, and economic pressures impinging on fishing in such a way as to make my purely biological explanation seem quite simplistic. This provided me with an important reason for examining not just the marine biological factors that influence fishing, but the human behavior patterns as well.

Most of my past research was written in a technical style that made it nearly incomprehensible to all but colleagues. The present research seems of sufficient general interest that I am seeking a wider audience in this book. So I have tried to write in a style that will interest colleagues, yet is sufficiently free of technical jargon so as not to discourage the layperson with an interest in the sea. I have tried to convey some of the excitement and occasional perplexity I felt as the work progressed. These are important ingredients in research, but ones which editorial conventions force us to ignore in more technical writing. Certain material has been placed in appendices so the general reader can ignore it if he chooses. Some of the more specialized material arising from this research is treated in greater detail in several technical publications (Johannes, 1977, 1978a; 1978b; 1980).

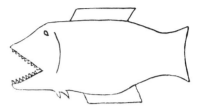

ACKNOWLEDGMENTS

My exceptional debt to Ngiraklang, second chief of Ngeremlengui and one of Micronesia's finest native scholars, will become
obvious as the reader progresses through this book. Patris Tachemaremacho, his father Patricio, and the rest of his family provided a
cheerful, endlessly patient fount of information on Tobian fishing.

One of the problems faced by this neophyte interviewer was that
of forgetting himself and asking long, rambling questions that
sometimes taxed interpreters. April Olkeril translated such questions (and some almost equally rambling answers) with a skill and
exactness that would put many professionals in the shade. Dale
Jenkins flavored his interpreting with rich insight and winning
cross-cultural humor. Beketat Maidraisau, Palau's first professional
marine biologist, provided much firsthand knowledge.

Other informants and interpreters in Palau included: from
Angaur; Mariek Remang, Moses Ngiramur and Abraham; from
Peleliu; Mark Mabel, Ngcheed, Mikel Mad, Nobuo Ngiradechal,
Ngirangesil, Uchau, and most importantly, Daelbai; from Koror;
Omelau, Ucherbelau Orrukem, Bena Sakuma, Ulou Brobesong,
Charlie Gibbons, Meruk Yalap, John Kochi, Alan Sied, Tewid
Boisek, Mongami, Cisco Uludong, Nancy Wong, Ngiromelau, Cap-
tain Seong Yoon Ho; from Ngeremlengui; Blau Skebong, Fumio
Kyota, Johanna Ongos, James Amboi Franz, Melemalt; from Ngiwal;
Sichang, Temol, Ignacio; from Geklau and Ngerechelong; the groups
of men I interviewed in the men's houses; from Ollei; Oshima,
Omelau Chereomel, Kloulechad; from Airai; Singeo Techong; from

Kayangel; Chief Rdechor; from Tobi; Zacharias Saimer, Kalisto Elias, Faustino Yaitowak; from Sonsorol; Mariano Carlos, Bernardo Tubito, Carlos William, and Pacifico Bemar.

People too numerous to list aided me in the pursuit of the marine lore of other Micronesian islands. I am particularly indebted to Juan Cepeda and Frank Cushing in Guam; Margie and Sam Falanruw and Issac Figir in Yap; Mike McCoy, Richard Croft, Ioanis Pretrick, Adam LeBen, and Al Milliken in Ponape; Rich Howell and Mario Henry in Truk; Jack Villagomez in the Northern Marianas, Bill Puleloa, Mike Trevor, and Pat Bryan in Majuro, and Teekabu Tikai and Mikaere Tekanu in Tarawa.

Others who helped out include Tosh Paulis, Jim McVey, Ted Tansy, Roger Pflum, Torao Sato, Lu Eldredge, Emilie Johnston, Sakei Moriss, Ronn Ronck, Howard Yoshiura, Renee Heyum, Douglas Faulkner, Darrell Gray, Gunter Reuning, Douglas Vann, and Roland Force. Bob and Hera Owen provided much expert advice and unstinting friendship. John Bardach helped shield me from the manic exactions of American academia while I wrote.

Chris, my wife, went, in one month, from completing her doctorate to scrubbing clothes on a washboard with good humor and a sense of adventure. My five-year-old son, Greg, offered tactful words of encouragement when my spearfishing was going badly and it looked as if we would have to eat canned mackerel again.

Portions of the manuscript were critically reviewed by Ted Hobson, William McFarland, Beketat Madraisau, Demei Otobed, Yoshihiko Sinoto, Robert Randall, Peter Black, John Munro, Ron Thresher, David Gibson, Jorg Imberger, and Richard Barkeley. The entire manuscript was reviewed by Gene Helfman, Bob Owen, Bill Wiebe, Malcolm Gordon, Craig Severance, and John Bardach.

Karen Straus, Gene Helfman, Robert Iverson, Brian Court, and Masachi Yamaguchi provided photographs. Dave Wright drew the fish, Liz Jefferson and Liz Corbin did the other illustrations.

The illustrations that appear above the headings throughout the book display some of the drawings of the 127 types of fishes made by an unidentified Palauan for A. Kramer (c. 1929).

These acknowledgments cannot begin to repay the many kindnesses received from these people and others in Micronesia, the United States, and Australia who helped make this work possible.

This work was supported by grants from the National Science Foundation and the John Simon Guggenheim Memorial Foundation.

R. E. J.

CSIRO
North Beach
Western Australia

C H A P T E R

INTRODUCTION 1

The Setting

Rising from the ocean floor 30,000 feet beneath the surface, an underwater mountain range stretches from Japan almost to New Guinea. At intervals along its 2,500 mile length peaks break the surface creating islands. Near its southern extremity an elongated cluster of about 340 such islands forms the archipelago of Palau.[1] They lie in the southwest corner of Micronesia, about 600 miles east of the Philippines and an equal distance north of West Irian. Within the group are high islands of both volcanic origin and uplifted limestone. In the north are two coral atolls. Clustered in the southern half of the archipelago are numerous small, almost unique, jungle-covered limestone cusps, each deeply undercut at sea level.

A barrier reef encircles most of the archipelago, creating a lagoon up to twelve miles wide, shallow in some areas, as much as 130 feet deep in others. Just outside this reef the bottom plunges steeply to abyssal depths. Much of the northern portion of the lagoon is occupied by the island of Babeldaob, which, with an area of 143 square miles, is the second largest in Micronesia. Four-fifths of its

1. Palauan place names vary considerably from one published source to another. And Palauans pronounce and spell some of their place names differently depending on whether they are communicating in English or Palauan. For example, "Palau" in English is *Belau* in Palauan. In this book I have used those names employed most frequently by English-speaking Palauans and Americans. Faulkner (1974) provides a map of Palau in which place names are given in Palauan.

ninety-eight mile shoreline is bounded by mangrove forest. The interior, with a maximum elevation of 794 feet, is covered with savannah and jungle. Only seven other islands are of significant size and only five of these, plus Babeldaob, are currently populated.

I chose to study fishing and marine lore in these islands for two reasons. First, Palauans have long been known as outstanding fishermen. Kubary (1895, p. 122) states, "As a result of the extremely favorable nature of the coasts of the Palau Islands and of the consequent abundance of sea creatures, and as a result also of the abundance of bamboo, of climbing plants with stalks that can be used for fiber, and of extremely good wood for vessels, the Palau islanders are in general excellent fishermen and, in this respect, occupy first place in Micronesia."

Second, these islands are characterized by exceptional marine biological diversity. The archipelago offers a greater variety of marine environments packed into a small area than any other place I know of. In addition to fringing, barrier, patch, and atoll reefs, there are extensive sandflats, mangrove swamps, a sizable estuary and taxonomically diverse sea grass beds. There are also about eighty small marine lakes and many sheltered marine coves that evolved geologically from such lakes.

A profusion of marine organisms inhabit these environments. Several years ago marine biologists collected thirteen new species and one new genus of fish on a Palauan reef in a two-hour period (Gene Helfman, pers comm.). The significance of this can be appreciated when it is considered that a total of only seventy-five to one hundred new species of fish are described annually from collections made throughout the world in marine and fresh waters combined (Cohen, 1970). The total number of fish species to be found in Palau is not accurately known but it undoubtedly approaches 1,000 (Jack Randall, pers. comm.)—several hundred more than can be found along the entire Atlantic, arctic, and Pacific coasts of Canada. Information gained about these species is not just of local significance. Because Palau is near the center of the vast marine biogeographic province known as the Indo-Pacific, many of the more common species found there are also found as far away as the Red Sea to the west and Hawaii to the east.

Palau has been inhabited for at least 2,000 years and perhaps for much longer (Osborne, 1966). The inhabitants are an athletic people of medium height. A wide range of contributors to the Palauan gene pool is manifest in their diverse skin tones and facial types. Kinky, curly, wavy, and straight hair are all common. Malays from Indonesia, Melanesians from New Guinea and some Polynesians from outlying Polynesian islands in Micronesia formed the basic stock. Europeans, Japanese, and Americans have made substantial genetic contributions in the past two centuries. Immigrants often slipped comfortably into Palauan ranks, for, more than other Micronesians, Palauans have sought new ideas and forms of expression through their contacts with the outside world. Once a newcomer settled, married a Palauan, and accepted the obligations of membership in Palauan society, he (there have been comparatively few female immigrants) was no longer referred to by his place of origin. He was now a Palauan.

The language is distinctive and difficult. It has many irregularities, perhaps reflecting the diverse origins of its people. It belongs to the western branch of the Austronesian tongue, with resemblances to languages of the Sulawesi (Celebes) area, a region also known for its skilled fishermen.

The sea is the Palauans' highway, recreation area, and, above all, their paramount source of *odoim* or animal food. Pigs, fruit bats, pigeons, jungle fowl, and, today, imported meat, provide an occasional meal, but sea food has traditionally been served at every meal. The chief traditional occupation of men is fishing, and there is no higher accolade than to be called a "real fisherman."[2] It is said jokingly by Palauans, but without exaggeration, that the subject of women is the only one that Palauan men discuss as avidly as they do fishing. Pride in fishing skill is matched by pride in knowledge of the ways of fish. Fish behavior is debated and analyzed endlessly by groups of older men as they sit crosslegged in the men's houses chewing betel nut.

Unlike many other Palauan activities, fishing cuts across boundaries of class and clan. Traditionally a chief is expected to be a good fisherman, but possesses no special authority and receives no special treatment while fishing. When Palauans fish, land-based protocol is suspended. Harsh criticism, or "words of the lagoon," *tekoi l'chei*[3] may be hurled by man or boy of any rank at anyone, chief included, whose efforts do not measure up on the fishing grounds. No one, irrespective of rank, may express offense at being scolded under such conditions. Thus has excellence in fishing been reinforced for centuries.

2. Reef gleaning—collecting small fish and invertebrates on the reef flat during low tides—was widely practiced by women. This activity is declining today. I twice interviewed women concerning their knowledge of the biology of gleaned species but was unable to elicit much information. I suspect that this was just bad luck and that more sustained efforts in this direction would have proven rewarding. Reef gleaning is a widespread subsistence activity in the tropical Pacific, yet it has received very little attention. In some Pacific islands this form of sea-harvesting is clearly still of great importance (e.g., Hill, 1978).

3. There is no single-word equivalent for *"chei"* in English. Strictly speaking it does not mean "lagoon" but rather the area stretching from the shore across reefs and lagoon to the point where the outer reef scope can no longer be seen from a canoe and the water turns from shallow-water green to oceanic blue.

Palauan society is based on a blend of pragmatic communalism and love of skillful, sometimes harsh political intrigue. Palauans approach life like a professional gambler approaches poker. An impassive demeanor masks an aggressive grasp of new ideas and opportunities. But in the twentieth century it has also come to hide a growing fear that the outside world is stacking the deck, for life has been changing erratically and uncontrollably as a consequence of foreign contact.

In the 1880s, Spain, which nominally controlled the islands, introduced Christianity. Germany bought the islands in 1899, built roads and conscripted native labor to mine phosphate and produce copra. Foreign pressures accelerated when the Japanese took over Micronesia in 1914 and made Palau the headquarters of their South Seas empire. They expropriated much of the land, greatly reduced the powers of the chiefs, and compelled the natives to work for them. They also viewed the islands as a partial solution to Japan's swelling homeland population. Before World War II as many as 24,000 Japanese lived in Palau—more than four times the total number of natives. And before the end of the war there were more than 50,000 Japanese military personnel stationed in the archipelago.

Koror, the district center, underwent development on a scale that is hard to imagine for the visitor today when he sees its dilapidated houses and rutted muddy streets. It was a "stylish metropolis, with . . . factories that manufactured soy sauce, beer and fireworks. The city had public baths, laundries, dressmakers, tailors, masseurs, barbershops, butcher shops, and drugstores. There were forty-one ice dealers, seventy-seven geisha girls, one fortune teller, and fifty-five restaurants, thirteen of them considered first class" (Kahn, 1966, p. 45). Exports included cloth, pottery, metal products, tuna, copra, rope, canned pineapple, cassava starch, glass, and machinery. This came to an end with World War II.

One of the bloodiest battles in the Pacific was fought in Palau. By the end of the war some Palauans were dead and many of the rest were very hungry. The jungles had been scorched or obliterated by American bombardments.[4] Gardens were untended and overgrown because their owners had been forced to flee. Almost all Palauan canoes had been destroyed by the Japanese to prevent Palauan contact with American invaders. The reefs had been severely over-fished in order to feed Japanese forces cut off from outside supplies.

After the war the United Nations awarded Palau to the United States as part of the United States Trust Territory of the Pacific Islands, which included most of that part of the Pacific known

---

4. "Koror, Malakal, Arakabesan and Babeldaob were bombed and rebombed. Much of this was in the nature of practice, utilizing targets so conveniently nearby" (*Military Geology of Palau Islands, Caroline Islands*, 1956).

geographically as Micronesia. Palauan gardens were replanted. The fisheries gradually recovered (but only temporarily as I will describe later). Beginning in the early 1960s heavy expenditures were made to build an education system patterned after the U.S. model. Palauan villagers sent growing numbers of their children to the new schools in Koror. After leaving school many found the slow pace of traditional life in the villages unappealing. So they chose to remain in the district center, despite serious unemployment coupled with rapidly accelerating rates of alcoholism, crime, and a variety of associated social problems.

Ngiraklang

When I arrived in 1974 more than 60 percent of the population resided in Koror. Typical diets included large quantities of imported flour, refined sugar, and canned fish. The average boy in Koror could not identify most of Palau's food fish. Neither in many cases could his father.

Under the circumstances it was obvious that most of my work could not be carried out in Koror but should be pursued with the older men in the outer villages. These villages, or "hamlets" as they are often called, are all on or near the coast. They are linked primarily by boat, although on Babeldaob seldom-used footpaths still mark eroded and overgrown roads built during the Japanese era. Here most men still fish and most women tend gardens of cassava, taro, and sweet potato.

The villagers of Ngeremlengui and Ollei were regarded by many, I was told, as Palau's best fishermen. Ngeremlengui, in addition, had the reputation of being both the most traditional and the most progressive village in Palau, a village governed by leaders who worked hard and intelligently to combine the best of Palauan tradition with the best of what the outside world has to offer. I decided to travel to Ngeremlengui.[5]

To reach this village of about 300 people, we travelled about ten miles by motor boat from Koror to the southwest coast of Babeldaob. From the lagoon the only sign of the village was a long stone dock that emerged from the mangrove forest, crossed the fringing reef, and ended at the edge of the lagoon. Nearby, a channel led into the mangroves, then became the tidal portion of a small river running between two curving walls of dense green foliage. A large, irridescent blue kingfisher flew up as we entered it. Several hundred yards inland the mangrove gave way to the lower part of the village, strung out along one bank of the river. The other side was bordered by a

5. The proper name of this village is Ngeremetengel. Ngeremlengui is, strictly speaking, the name of the municipality in which Ngeremetengel is situated. The village, however, is usually referred to in Palau as Ngeremlengui, and I follow that custom throughout this book.

high sheer cliff from which tiny waterfalls sprayed into the river. Several miles beyond the village a limestone outcrop was visible rising several hundred feet out of jungle and savannah.

Along the wide village path running beside the river, young boys played in small bands, staring at us as we approached. Several men lounged on a bamboo bench under a large tree, talking. Nearby in the men's house older men sat crosslegged discussing village business. As we disembarked, a stooped but sturdy and alert-looking old man came out of the men's house to greet us. He bore the name Ngiraklang, hereditary title of the second chief of Ngeremlengui. He invited us to sit in the shade of his porch, and there listened attentively as the Palauan Marine Resources Division officer who brought me explained the reason for my visit. He agreed to my request; there was an unoccupied house nearby in which my wife, young son, and I could live.

A few days later, after we moved into the village, I approached Ngiraklang to find out who among the men in the village I might profitably interview. "I do some fishing," he said. "You may start with me if you wish." In the evenings we talked through an interpreter by the light of a pressure lamp. Occasionally we went fishing in his boat when village business allowed him. Gradually over the next weeks I came to recognize the extraordinary understatement in his remark, "I do some fishing," and my luck in finding him among 13,000 Palauans at the beginning of my work. For it was largely because of his knowledge and leadership that Ngeremlengui's fishermen had earned their reputation. He was Palau's most knowledgeable fisherman. (Once later, when interviewing a group of fishermen in a village twenty miles away in a location where fishing conditions differ substantially from those around Ngeremlengui, I was told, "We are sorry we cannot answer some of your questions. You should ask the fisherman of Ngeremlengui called Ngiraklang. He understands more about the fishing here than we do").

He was born in 1894. Beginning in his teens he studied fishing avidly, first learning everything his uncle could teach him, then turning to other men to add to his knowledge. Not willing to accept things without proof, he tested what he was told and thereby came to discard or modify a number of beliefs held widely by other fishermen. His original brash intention as a youth, he confided, had been to become the best fisherman in Palau. When the older men gathered and talked of fishing, he listened and hoped some day to know more about fishing than anyone else in the men's house. By middle age when he had mastered most traditional fishing knowledge, his curiosity had outrun his ambition. He had become fascinated by fish for their own sake, independent of their role as Palau's main source of animal protein, and of his ability to catch them. He found himself observing the habits even of small fishes to which others paid little attention because of their lack of food value.

His curiosity about the natural world was not limited to the sea, however. He also mastered the traditional body of knowledge concerning blooming and fruiting cycles of Palau's important plants.[6] Then once again his curiosity took him beyond the practical and he took to examining and memorizing the blooming cycles of the weeds along the village path. He studied the nesting patterns and feeding behavior of birds and the rhythms of insect abundance.

He carried out experiments to further his knowledge. He transplanted giant clams from their normal habitat in shallow water to the bottom of the lagoon and checked them periodically for three years to see if they would grow. (They didn't.) He transplanted trees closer to his house so he could more readily observe their blooming cycles. He set up hypotheses and tested them. In one area where fish were mildly toxic, he incubated untainted fish in the water to see if it was the water that was the source of the contamination, doing what a biologist would call a bioassay. He was, in short, a self-made scientist—with one-half year of formal schooling, in carpentry—in the 1930s.

Eager to master things outside his culture (he was the first man in Ngeremlengui to have a motor boat), he was nonetheless an authority on Palauan culture. "We have been collecting and writing down Palauan stories and legends," a Palauan in Koror told me. "And after we get a new story from one of the old men we check with Ngiraklang to see if it is correct. He is our final authority."

As our interviews progressed I learned that his knowledge of reef fishes was phenomenal. Perhaps most impressive was the fact that he knew the lunar spawning cycles of several times as many species of fish as had been described in the scientific literature for the entire world. (This knowledge is described in Chapter 3 and appendix A.) He was a stickler for detail. One night during an early interview, I apologized for asking him so many questions concerning what I felt might seem to him to be insignificant details. "You need not apologize," he said. "Many important things can be learned through bringing together small details." Often he volunteered additional details when he felt I had neglected to pursue a subject as far as was profitable.

At first I wondered why this busy and important man should be willing to devote so much time to the education of a stranger, particularly when his own time was running out. One day he gave me the answer: "I am glad you have come," he said, "because through you I can leave my footprints in this world before I move on to the next." He was too courteous, I believe, to have said this if he

6. Some of this information has been recorded by Klee (1972).

had known the inadequacy I felt in having been given a respon-sibility I was not capable of discharging effectively. One can easily learn and record knowledge of fish behavior and fishing. But learning the *skills* necessary to become a good fisherman takes years of observation and practice. (As time went on I managed to become a tolerably good underwater spearfisherman, but remained close to hopeless, by Palauan standards, at other types of fishing.) Such skill cannot be transmitted adequately in a book. However it was no longer being transmitted effectively by traditional means either as the young abandoned the villages. So this book represents the best I can do to help preserve the fading footprints of Ngiraklang and the other expert fishermen of the Palau district.

Interviewing

Eventually I interviewed about thirty-five Palauan fishermen singly, plus about thirty more in groups throughout the archipelago. (Later I had opportunities to interview fishermen in a variety of other parts of Micronesia as well, and some of the information I obtained from them is also included in this book.) How reliable was the information I obtained?

Anthropologists I had consulted before going to Palau warned me repeatedly that Pacific island informants sometimes framed their answers to questions with a greater regard for pleasing the inter-viewer than for the truth. Answers might be distorted to conform to the interviewer's perceived opinions, or invented in cases where the real answer was not known. But a biologist asking questions about organisms and environments with which he has some familiarity has an advantage over an anthropologist asking questions about an unfamiliar culture. Palauan fishermen knew volumes about the inhabitants of their reefs and lagoon that I did not. Nonetheless I, as a marine biologist, knew a lot about some of these subjects too. This enabled me to test the reliability of my informants by asking of them two types of questions.

First there were questions to which I already knew the answers. To such questions I almost always received either the correct answer or an admission of ignorance. Second were questions that I felt sounded plausible, but which I knew that the fisherman could not possibly answer. In virtually every instance when I asked this kind of question I received an immediate "I don't know." The only signifi-cant departure from this pattern was with a single fisherman from Yap, a nearby island group, whose fertile imagination, I discovered later, was well known. I received no deference during my interviews. I was often corrected and sometimes gently scolded for having forgotten something I had been told earlier. This was encouraging, but it was also perplexing. This was not what I had been led to expect. Why were things going so well?

Later an anthropologist friend to whom I put this question came up with the probable explanation. It had to do with the difference between the kinds of questions I was asking and those often asked by anthropologists. When a Pacific islander is asked about his eating habits, his kinship system, or his sexual customs, he is liable not to see the point of the question. He seldom analyzes his culture, he simply lives it. Often he may not be able to verbalize the correct answers even though he may "know" them insofar as they are an unconscious part of his culture. (How many Westerners, if the tables were turned, could tell an interviewer how much coffee he drank yesterday, how many aunts his mother has, or why, in America, breast size has a strong influence on bride selection?)

In contrast, the fishermen could easily perceive the purposes of most of my questions. My interests were similar to theirs. More than once lying on the floor of a men's house I awoke at dawn to see a group of fishermen sitting in the same positions they had occupied eight hours earlier when I had left off talking with them in order to sleep. Their sleeping mats had never been unrolled. They were still intently debating questions that had arisen during the previous night's interview. How well the men of the villages knew the answers to such questions was a measure of how good they were as fishermen, and they were proud of their knowledge. Also, like Ngiraklang, they realized that much of this knowledge was on the verge of being lost unless it was written down, and they wanted it recorded correctly.[7] (Additional inducements to accuracy were perhaps superfluous. But the fishermen also knew that I was asking the same questions all over Palau, sometimes quoting their answers to friends or competitors, and that I routinely spent several hours a day on the fishing grounds checking on what they had told me.)

More good fortune came in the form of unexpected access to expert fishermen of a different culture, the inhabitants of three tiny islands south of Palau. Faced with different and less congenial fishing conditions than Palauans, they nonetheless depended on the sea for their food to an even greater degree. How they have adapted to these conditions, and how, in consequence, their fishing methods and knowledge of the sea differ from Palauans' are the subjects of a later section of this book.

7. The other great body of knowledge of which Pacific islanders are justifiably proud concerns navigation at sea. Lewis (1972, p. 10) who studied this lore noted:
> Our inquiry took place at a period in history when everyone realized that the ancient lore was on the verge of extinction without a trace—unless recorded in writing. . . . The navigators proved uniformly anxious to make sure that everything they expounded and demonstrated was grasped correctly. . . . The navigators' extremely responsible approach to the knowledge of which they were custodians virtually precluded intentional deception. I have no hesitation in making these assertions even though I am well acquainted with the Pacific islanders' sense of humor and love of exaggerating personal achievements.

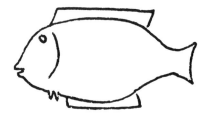

CHAPTER

PALAUAN FISHING METHODS

2

Palauan fishermen have names for more than 300 different species of fishes and distinguish readily between species differing only subtly in appearance (Helfman and Randall, 1973).[1] This ability was often in evidence when, during interviews, I leafed through an ichthyology book in search of a picture of a particular species. Each pair of pages had about twenty-five small black-and-white pictures of closely related species. I was familiar with these pages, whereas the fishermen were not. Yet in many instances they would point to the right picture before I spotted it. This was all the more impressive because the fishermen were facing me and were thus looking at the pictures upside down. This intimate familiarity with local fish is also manifest in the many different fishing techniques used by Palauans—techniques that take into account differences in anatomy, behavior, and habitat of many different species. Traditional Palauan methods have already been dealt with by Kubary (1895) and Kramer (1929). I will describe a few of the more noteworthy ones here before passing on to contemporary methods.

1. Kubary (1895) similarly noted "how excellently" Palauans could draw "the shapes and characteristics" of different fish species.

Throwing Spears

Until recently fish were commonly hunted with throwing spears in six to eight inches of water on the reef flat.[2] Small mullet or *uluu (Chelon vaigiensis)* forming loose schools of about twelve individuals were often the target. When the school was chased it eventually broke up. When a man spotted a school of fish he whistled or shouted to draw it to the attention of the others. Each man picked a fish and chased it, trying to force it into shallower water. Because the fish tried simultaneously to get into deeper water, the resulting chase often paralleled shore. Usually after the seventh jump, according to fishermen, the fish would slow down and present an easier target. At this point the fisherman tried to get alongside the tired fish and throw the multiple-pronged *taod*[3] spear on a low, flat trajectory, aiming just in front of the fish. This was often done on the run, the spear thrown from perhaps ten feet away with a sidearm toss.

If rabbitfish were sighted, an attempt was made to get all the fishermen between the school and deep water and drive the school into the shallows where its members could easily be speared. One escape response of the fish was to take shelter alongside or under a log or boat. Once under such a floating object the fish were harder to spear because, as the object moved, they moved with it. Often a rabbitfish would head straight for its pursuer and hide in the silt stirred up around his feet. Mullet sometimes used the same tactic. A single-pronged *klibiskang* spear was preferred for rabbitfish; they presented easier, slower moving targets than mullet and, once impaled, could be pushed up the shaft and stored on the spear so that the fisherman could work faster.

Spears were (and still are) also used for subduing large fish caught with hook and line. Before being boated the fish are killed by thrusting the spear into the brain. Traditionally if a particularly large fish, such as a billfish, proved especially hard to handle, a special basket was first slid down the line, coming to rest over its head. This is said to have a marked calming effect on the fish.[4]

---

2. This account comes from information provided by Gene Helfman. In 1968 Helfman saw this activity commonly. By 1974, when I arrived, fishing in this manner had almost died out and I never witnessed it.

3. Palauan names for various items of fishing gear, and for fish, vary somewhat from place to place in the archipelago. I have chosen to use the names that were used most commonly by fishermen I interviewed.

4. Similarly when large parrotfish are removed from nets, fishermen tuck the fishes' heads under their arms before carrying them to shore; covering their eyes, it has been found, causes them to cease struggling.

Using Marine Animals to Catch Marine Animals

When the skin of a sea cucumber, *Holothuria atra (choas)*, is rubbed it emits a red liquid containing a nerve toxin. It was used by Pacific islanders to kill fish in shallow pools at low tide.[5] Also, when introduced into an octopus lair it drove the occupant out into the open where it could be speared. In Palau the toxin was also employed to paralyze large edible sea anemones used as food. Once exposed to the substance the anemone would neither retract into its hole nor sting the fisherman as he dug it from the reef.

Mantis shrimps were lured to the tops of their burrows with bait, then snagged and dragged out with a device employing the claw (second maxilliped) of another member of the same species. This claw was lashed to a stick, and the bait was secured above it. When the device was lowered into the hole the claw folded up on itself. The shrimp moved up past the claw to get the bait. As the stick was pulled from the hole the claw opened like a partially open jacknife, its "blade" preventing the shrimp from retreating.

As elsewhere in Oceania fishhooks were made from the shells of hawksbill turtles (green turtle shell is too weak) and of various molluscs. Although no record exists of how these hooks were made, it is possible that they were shaped, as on some other Pacific islands, using coral files, and finished using the abrasive skin of rays.[6] Throwing spears were sometimes tipped with the spines of stingrays.

The Leaf Sweep

Using a rope festooned with leaves to herd and capture fish might sound as if it were guaranteed to fail. But the leaf sweep was used traditionally throughout much of Oceania, from Hawaii to Palau, because of its simplicity and effectiveness. It is an example of the island fisherman's use of applied fish psychology. Called *hukilau* in Hawaii and *ruul* in Palau, the device resembles a giant green Christmas tree garland. It has been all but superseded by nets in much of Oceania and I saw it being made and used in Palau only once.

Although fish can easily swim physically through or under a leaf sweep, it forms an effective psychological barrier for most species.

5. Whereas sea cucumbers were used traditionally in Oceania as a fish poison (e.g., Frey, 1951; Smith, 1947) biologists established their toxicity only in the 1950s. Some of the first research of this nature was carried out, coincidentally, in Palau by Yamanouchi (1955), who found *Holothuria atra* to be one of the most toxic species present. Abe (1938) also lists four plants that were used as fish poisons. As far as I could determine only one, *Derris elliptica*, is still used occasionally today. The Japanese imported this species, a more potent variety than the indigenous *Derris trifoliata*, and raised it commercially in Palau for the manufacture of insecticides. They also encouraged its use by Palauans to secure fish for them.

6. Ray skins were used until early in this century in Palau as an equivalent of sandpaper in smoothing canoes and wooden utensils.

Ony a few, such as the rabbitfish, *Siganus canaliculatus* (*meas*), who live in seagrass beds and are accustomed to navigating leafy underwater jungles, will normally escape by swimming through or under the *ruul*.

It takes the better part of a day for a dozen Palauans to make several hundred yards of *ruul* by weaving palm fronds around a woody vine or rope. The completed *ruul* is piled onto two bamboo rafts and transported onto the reef flat on a falling tide. The two rafts are slowly pulled apart and the *ruul* is paid out into the water. The rafts are moved in opposing arcs, resulting in the deployment of the *ruul* in a circle. When the circle is complete, the fishermen then begin to gather in the *ruul*, piling it back onto the rafts while pushing them together toward the far side of the circle. The *ruul* can be used in coral-studded areas where a beach seine would snag hopelessly; as it moves the stiff leaves simply brush across the corals.

As the circle of leaves shrinks, swirls of water begin to betray disturbed fish, and fishermen clustering around the edge start shouting excitedly to one another. When the circle is a few meters in diameter, one of the older fishermen, who acts as a kind of foreman, calls a halt to the hauling. Fishermen begin spearing fish out of the milling aggregation. It is a crucial time, for if the fishermen create too much commotion the fish may "stampede" and escape.

A variation on this strategy is to interpose a special wooden fish trap (*semael*) or a net (*osel*) between the two sections of the *ruul* and herd the fish into it. Fish can be kept in a *semael* for several days until needed.

Shark Fishing

Rich and extensive reefs and lagoons provided Palauans with more than enough seafood. Traditionally, therefore, fishermen seldom ventured much beyond the outer reef slope (Kramer, 1929; Kubary, 1895).[7] But one type of offshore fishing, *oungeuaol*, was practiced by a few prestigious specialists. On special occasions at the request of the chiefs these men fished the open ocean for several species of sharks.[8] During the season of the northeast trades,

7. It is possible that the residents of Angaur, the isolated southern-most island of Palau, were once accomplished open-ocean sailors because their island has only a limited and poorly protected reef. But old accounts of fishing in Palau omit significant mention of either Angaur or the northernmost atoll of Kayangel. In fact the observations of both Kubary (1895) and Kramer (1929) were restricted largely to fishing along the eastern coast of Babeldaob, which both authors mistakenly assumed to be representative of Palau as a whole.

8. I asked Ngiraklang why *oungeuaol* fishermen went so far from shore to catch sharks when they were so abundant nearer to shore around channel mouths and along the outer reef slope. He said that the species encountered offshore were generally less dangerous than those caught in shallow water. Melanesian natives volunteered the same information to Codrington (1891).

oungeuaol fishermen sailed up to ten miles off the east coast of Babeldaob looking for floating driftwood around which they knew sharks often congregated. Flying fish were caught and used to lure the sharks close to the canoe where they were caught with a noose made from hibiscus fiber. This technique was widespread in Oceania (see further discussion, p. 94) and is still in use today in parts of Melanesia. It was devised in response to the fact that sharks are liable to cut through a conventional bait-fishing line with their teeth.

Whereas sharks used to be esteemed as food by Palauans they are not popular today. Although they are common around Palau's reefs I saw only one brought in by fishermen. Their prestige, according to Ngiraklang, was related not to their flavor, but to the romance and danger associated with their capture. (The *oungeuaol* fisherman had a special tattoo on his wrist. When holding out a flying fish to entice a shark to swim through the noose, he was not supposed to let go of the bait until the tip of the shark's snout reached the tattoo).

Oungeuaol fishing died out around the turn of the century (Kubary, 1895), having been discouraged by Spanish and German colonists. For not only was it dangerous, but months of labor were also involved in preparing for a special celebration associated with the completion of the trip. *Miich* nuts, tropical almonds of the tree *Terminalia catappa*, were pounded and mixed with coconut syrup and starch in order to create shark sculptures as much as three feet high and six to eight feet long. Some was eventually eaten but much was wasted.

Other elaborate rituals often attended fishing and are described by Kubary (1895) and Kramer (1929). Sexual intercourse was forbidden prior to a fishing expedition.[9] This taboo, like most of the rest, has died out, and fishermen say laughingly that they are happy to be rid of it.

Gorges

Gorges of several types were sometimes used instead of hooks for dropline fishing and trolling (Kubary, 1895). These were straight

---

9. This prohibition was widespread in the Pacific islands and must be a custom of great antiquity. Presumably it once served an important purpose, but I have found no published explanation for it. (Reichel-Dolmatoff [1974] provides a remarkable metaphysical explanation for a similar prohibition among hunters and fishermen of the Tukano Indians in the Amazon.) Fishermen from the island of Yap, 300 miles north of Palau, maintain that the sea is a jealous woman. If she smells the presence of another woman on a fisherman she will withhold her favors and fishing will be poor. These enterprising men have devised a way to circumvent this amatory impediment. A certain plant, when rubbed on a man, eliminates the telltale odor and fools the goddess of the sea.

sticks pointed at both ends and tied in the middle to the line. They were baited and deployed in such a way as to lie parallel to the fishing line. When a fish swallowed the bait and tension was applied to the line the gorge turned sideways and lodged in the stomach or throat. Although gorges are generally not as effective as well-made hooks, they coexisted with fishhooks in Palau because they were much easier to make. A good turtle shell hook took two or three days to make, whereas a gorge could be made in minutes. One Palauan in his nineties made for me some gorges like those he remembered having used in his youth.

Fish Weirs

Stone or wooden fish weirs were built on the reef flats to trap fish on receding tides. The maintenance of these structures required considerable labor. They have fallen into disrepair in this century, being replaced in function by the less labor-intensive *kesokes* net (see later in this chapter). In 1975 I heard of only one stone fish trap, near Geklau, which, although not maintained, still afforded a few fish occasionally to the owners. Vestiges of another, just north of the dock at Ngeremlengui, were visible in 1976. It had ceased to catch fish in the 1950s.

Modern Methods

Fishing, like many aspects of Palauan culture, has changed considerably in this century. Motor boats have almost completely replaced dugout canoes. Imported machine-made nylon nets have replaced locally made varieties. Fishing techniques have been modified and reduced in number. The elaborate religious rituals once used to placate the gods of weather and fishing have almost completely disappeared.

The classical cultural anthropologist might say that fishing today in Palau is a pale shadow of what it once was. But the biologist interested in what can be learned from Palauans concerning the biology of reef fish can learn much that was unknown to Palauans fifty years ago. For in Palau, as in many other Pacific islands, there was a recent period of several decades when old and new ways of fishing overlapped considerably, producing a degree of understanding of marine life that was probably never before equalled (and may not soon be equaled again).

While older Palauan fishermen were still teaching the young traditional fishing knowledge accumulated over centuries, modern technology was simultaneously expanding fishermen's horizons. Remote and seldom-visited fishing areas became within easy reach by motorboat. Much of what Palauans know today about fish outside the reef, for example, was first learned during this period. A number of common offshore species such as skipjack and yellowfin tuna are known today by Japanese names—a legacy of Palauan employment in Japanese fishing activities beyond the reef.

A revolution in the study of underwater life began at the turn of the century when diving goggles were imported by German traders to facilitate diving for pearl shell. In the late 1940s underwater spearguns were also introduced. A few years later underwater flashlights became available, facilitating nighttime underwater spearfishing. This equipment enabled fishermen to exploit habitats and harvest species to which they had little previous access. Species that would not take hooks, or were too powerful, or distributed in waters too deep to be taken readily in nets, now became available. And a fisherman could now easily observe the fish he sought while he was actually submerged in the clear waters of their habitat. Even at night (when the behavior of many species changes radically) he could watch them at close quarters. Such developments led inevitably to a heightened understanding of fish behavior. As Ngiraklang stated, "We perhaps cannot name as many fishes as our forefathers, but the fishermen of my generation know more about their habits." (Why much of this information is being lost as Ngiraklang's generation passes will be discussed in later chapters.)

Today in Palau underwater "fishing guns" and stationary barrier nets are the two most productive and extensively used fishing devices. Gillnets, droplines, trolling lines, throwing spears, and portable fish traps account for most of the rest of the catch.

Stationary Barrier Nets

The traditional movable *ruul*, or leaf sweep, described earlier has given way almost completely within the past few years to an imported stationary barrier net called *kesokes*. In keeping with the shift from collective to individual labor that often follows Western contact in Oceania, it can be employed by a single individual, whereas the *ruul* requires many people to operate. It is usually several hundred yards long, from two to five feet deep and has a stretched mesh of from one to two and one-half inches. It is composed of detachable segments and its length can quickly be adjusted to particular needs by adding or subtracting segments.

At Ngeremlengui the net segments are placed on a bamboo raft, or *brer*, and poled out onto the reef flat. The segments are tied together and the net is set on the reef flat in a V shape on a falling spring tide. The orientation of the V is important; a good set requires that the fisherman know the regular pathways the fish take as they leave the reef flat for deeper water as the tide falls (see chapter 3). The apex of the *kesokes* usually lies in the deepest water in the area.

It is sometimes set on a portion of the reef flat which protrudes into deep water because some species of fish tend to leave the reef at the tips of these promontories. As the tide falls the fish move toward deeper water in the apex. By the time the tide has dropped so far that the water is uncomfortably shallow for fish in the apex, the reef flat is dry in the mouth of the net, so they are trapped. As the fish flop around in the receding water they are speared, often by small boys who get their first training here in the use of the hand-thrown, three-pointed spear or *taod*.

As many as three sets per trip may be made during the ebb of an extreme spring tide. At this time speed is of the essence; the fisherman cannot afford to waste time picking up fish while the tide is dropping. So, after the fish have concentrated in the apex of the first set, he simply surrounds them with one or two segments of the net and moves to another spot lower on the reef flat to reset the rest of the net. Once the reef flat is dry and no more sets can be made he returns to gather the encircled fish.

Kesokes nets are set at night as well as during the day. Poles driven into the reef mark the correct spot to place the apex of the net so that the fisherman can place it properly even during the dark of the moon.

Such nets have several advantages over the *ruul* although their cost is much greater. The rabbitfish, *Siganus canaliculatus* or *meas*, cannot penetrate the *kesokes* as they can the *ruul*. (After the introduction of the *kesokes* this fish became the most important single species in the catch of the fishermen of Ngeremlengui.) *Kesokes* nets are lighter, less bulky and thus easier to deploy than the *ruul*. They also last much longer; a *ruul* lasts only a week or two. The owner of a *kesokes* net can catch many fish with comparatively little labor, but the method has one serious drawback. Small fish that are otherwise capable of swimming through the meshes are often entrained with the larger fish and die when exposed as the tide recedes.

Underwater Spearfishing

Palauans traditionally used long, hand-held spears underwater. But before the introduction of diving goggles this method was not very productive or popular because of the poor acuity underwater of the unprotected human eye. But where fish were abundant and not very wary, such spears, used with goggles, were productive in the hands of skilled spearfishermen.

With a hand-held spear the fisherman would dive to the bottom and lie flat and still, holding onto a coral head to help him stay down. He could depend on certain species such as ring-tailed surgeonfish (*mesekuuk*) to approach him out of apparent curiosity (see also Clark, 1953). Once a fish came within range the diver would impale it with a quick thrust.

An underwater speargun with a detachable shaft propelled by rubber was reportedly first introduced to Palau in the late 1940s by an Indonesian fisherman who fell ill and left his ship to recuperate in the Koror hospital. Because of the greater power, range, and speed of this device, Palauans adopted it eagerly and many are now highly skilled in its use.

Today most village men have at least one fishing gun. The stock is made from local wood, the trigger mechanism from stainless steel scraps, the propellant from surgical rubber tubing, and the barbed shaft of special spring steel reinforcing rod, often attached to the gun with a strong cord. From butt to speartip the loaded gun varies in length from about three and one-half to more than nine feet.

Long guns are generally used during the day when most fish are not easily approached closely. Such guns are usually made of wood that is of neutral or slightly positive buoyancy. This makes pivoting with them in the water easier and prevents their loss in deep water if for some reason they are dropped.

Fish are stalked cautiously. The Palauan, like other Pacific islanders, submerges to go after a fish in a manner different from that of most Western spearfishermen. Rather than upending and going down head first, he pulls in his abdomen, pulls his arms tight to his chest and sinks feet first. (The greater buoyancy of a fatter fisherman makes this maneuver difficult and he is subjected to good-natured joking about it.) The purpose of this strategy is to approach the fish with a minimum of disturbance. When the diver gets ready to shoot, his body and the speargun are held in a straight line so as to present the minimum visual area to the fish, thereby reducing the chances of alarming it. (This is just the opposite of the behavior of some reef fishes, such as lionfish and some surgeonfish, when threatened. They elevate their fins so as to make them appear larger and, presumably, harder to swallow.) Palauans prefer nonbuoyant fins and sometimes tie small lead weights to them. This is to counteract the tendency for a diver's feet to float slowly upward while he is stalking a fish horizontally.

The effective range of the longest gun is about ten feet beyond the spear tip, or about seventeen feet beyond the trigger hand of the diver. Often attached to the butt of the gun is a forty- to fifty-foot length of line. Polypropylene line is preferred because it floats and is thus less liable to hang up on corals as it drags behind the diver. The other end of the line is attached to a float—usually a five-foot length of bamboo four or five inches in diameter. Attached to this float is a separate cord eight to twelve feet long, the free end of which terminates in a copper spike about the size of a pencil. This line or *iengel* is used for stringing fish as they are caught. The spike facilitates threading the line through the mouth, gills, or eyes of the fish. By dragging the fish behind him rather than tying them to his waist the diver is not only freer in his movements but is also less liable to be threatened by sharks attracted to his bleeding fish.

The bamboo float serves a second purpose. When a fish or a turtle is speared that is strong enough to break the line attaching the spear to the gunstock, or to drag and hold the diver down, he lets go of the stock. His prey swims off towing the gun, rope, and float, gradually exhausting its strength without putting undue strain on the cord attached to the spear. He retrieves prey and gear when the float pops up from beneath the surface signaling the end of the struggle—which may take an hour or more if a large turtle has been speared.

Sea turtles are usually speared in a front flipper. A turtle thus speared is forced to swim in circles (only the front flippers are used for propulsion in these species) and is unable to fight effectively. Fish are usually speared from one side, or from above and behind. The diver aims at or just behind the head; the spear holds better when imbedded in the bones of the skull. If the shot is perfect it hits the brain or severs the spinal cord just behind the brain and the fish dies instantly.

Divers exploit various behavioral traits of their prey. If a struggling jack or large surgeonfish is left on the spear, other members of the school from which it was shot will often crowd around it, providing easy targets for other spears. Rudderfishes, *Kyphosus cinerascens* and *K. vaigiensis* (*komud*) lose their customary wariness and will approach the diver and eat the partly digested algal fragments released when a spear penetrates the stomach of one of their school.

Sounds made by various reef fishes are sometimes employed by the underwater spearfisherman to locate or attract prey. Several species of angelfish bear the onomatopoetic Palauan name, *ngemngumk*. When said quickly with an emphasis on the vowels, the word reproduces the percussive grunting sound these fish make when alarmed. The squirrelfish, *Adioryx spinifer* (*desachel*) make a similar, although softer alarm sound. Both squirrelfish and angelfish often hide when alarmed, but these sounds advertise their presence and general location to the spearfisherman.

Some jacks are readily attracted to the fisherman if he makes a glottal grunting sound similar to the noise the fish themselves sometimes make by grinding their pharyngeal teeth (e.g., Moulton,

1958).[10] Some jacks are also attracted to a diver if he expels bubbles noisily from behind pursed lips as if blowing on a trumpet. (This technique worked very well for me with certain species, providing I was at or below the level of the fish in the water. The jacks would change course abruptly and swim straight toward my face, sometimes approaching so fast that I was unable to pull my long speargun back and take aim before the fish were at such close range that it was impossible to shoot.)

Some species, particularly rudderfish *(komud)* are said to become alarmed if a diver presents his face to them. This is believed to be a reaction either to the diver's eyes or to light glinting off his faceplate. In any event such species will often not allow the diver within spearing distance if he keeps looking at them. (Native divers in Papua New Guinea have made the same observation [Gaigo, 1977].) Accordingly the diver faces the bottom, peeking awkwardly upward occasionally to keep track of the fish's movements as he swims slowly toward it. (I was always too eager to watch the fish I was stalking to test this technique adequately.)

Palauans say that large barracuda, dogtooth tuna, or large jacks should never be speared head on. Sometimes these fish will charge straight ahead the instant the spear is fired and may inadvertently drive the butt of the spear into the diver. Most other injured fish, they say, generally flee down current. (This seems adaptive because the trail of blood left by an injured fish fleeing upstream would enable predators to trace it more easily.)

Octopus are highly valued as food and bait, and their well-known ability to camouflage themselves makes them hard to spot on the reef. Divers seeking octopus look for various signs of their presence. Piles of broken crab or clam shells often indicate the presence of an octopus burrow. As an octopus crawls along the bottom its suckers stick to pebbles and turn some of them over. Certain pink coralline algae that grow best at low light levels are often found growing on the bottom of such pebbles. A trail of pink pebbles is thus a sign that an octopus has passed.

If an octopus is too far back in its hole to be grabbed or speared, divers attempt to force it out. Moray eels are among octopus' worst enemies, and a dead moray thrust into an octopus lair often brings the occupant flying out. The neurotoxic secretion of a sea cucumber,

10. Beaglehole and Beaglehole (1938), Kayser (1936), and Bagnis et al. (1972) also mention the use of sounds by underwater spearfishermen to attract fish in other Pacific island groups. I heard accounts of one Palauan fisherman who coaxed groupers out of their caves and into shooting range by holding small pieces of coral rubble in his hand and grinding them together in front of the hole. An American witness said that the groupers came out aggressively as if defending their territory in response. I was not able to get this technique to work on the few occasions I tried it.

described earlier, has the same effect. Once captured the octopus is often killed by biting it between the eyes to destroy its brain.

Two divers often work together over sand flats containing occasional coral patches. When a large fish is spotted one diver pursues it slowly toward a coral patch. The fish usually swims around the edge of the patch staying on the opposite side from its pursuer. Meanwhile the second diver hides behind a coral in its path and spears it as it passes by. A variation of this can be carried out by a single diver if the coral patch is not too large. Once he has the fish swimming away from him around the patch, he reverses direction and spears it as it meets him.

A few Palauan fishermen have had the opportunity to use SCUBA while spearfishing, enabling them to go deeper after fish than has been customary. Two of these men told me independently of observing interesting behavior in fish they were chasing down the reef slope into water 100 feet or more deep. The fish suddenly stopped and would go no deeper. On arriving at this depth the fishermen discovered that they had hit a thermocline—the water abruptly became much colder. Apparently the thermocline represents a barrier so pronounced for some reef fishes that they will not penetrate it even when being chased. (This is also true of tuna [see p. 97].)

Sharks are common in Palau and it is a routine experience to be approached by one or more while spearfishing in channels through the fringing reef or over the outer reef slope. Palauan divers often proceed about their business with what may seem to be a total lack of concern while sharks patrol the immediate area. Many tales of "fearless" Pacific island divers are probably based on observing such behavior. But the appearance is deceiving. Changes in a shark's behavior that may not even be noticed by less-experienced divers signal to Palauans that it is growing less cautious and may soon become aggressive. At this point the diver may retreat, or at least stay close to his boat and keep a watchful eye on the shark.

If more than three sharks remain nearby, a diver becomes wary even before they show signs of imminent aggression, and sometimes leaves the water, particularly if the objects of his concern are lemon sharks, *Negaprion acutidens (metal)* or grey sharks, *Carcharhinus amblyrhynchos (mederart)*. Members of a group of sharks are said to be considerably less cautious than isolated individuals and may shift much faster from exploratory to aggressive behavior (see also Hobson, 1963; Myrberg, 1978).

Palauans are more wary of small sharks than large ones, saying that smaller individuals tend to be less cautious and more aggressive. Others have made similar observations (e.g., Bagnis, et al., 1972; Becke, 1905; Hobson, 1963; Myrberg, 1978).

The more people there are in the water the less likely sharks are to attack, according to divers. So in areas where sharks are especially abundant, such as off the southeast coast of Peleliu, divers prefer to work in teams. Here at the reef dropoff divers sometimes form queues by an anchored fish stringer. One man dives down, spears a fish, and heads back to the surface. Several grey or lemon sharks often follow. At this instant the next man in line dives toward the sharks, hooting and feinting at them, causing them to withdraw a few yards. The man with the fish on his spear removes it and kills it by driving the spear through its brain either through the top of the head or through an eye socket. This is done so as to cut short its struggles and reduce the attraction of sharks. He then hands the fish to the stringer-tender. Meanwhile the man who had diverted the sharks is now on his way down to spear the next fish, and the man behind him is getting ready for the next diversionary feint.

Fish that vibrate rapidly and audibly on the spear as they struggle, such as jacks and surgeonfish, act as exceptionally effective shark attractants. In an area where no sharks have been seen during a dive, several may suddenly materialize seconds after such a fish has been speared. (Sharks may travel hundreds of meters in tens of seconds in response to sounds [Myrberg et al., 1976] and are some-times attracted toward a diver merely by the sound of a discharging speargun [Nelson and Johnson, 1976].) The noises made by fish that struggle only lethargically on the spear, such as groupers, are less attractive to sharks. (Banner [1972] recently demonstrated this experimentally, finding that the relative attractiveness to young lemon sharks of sounds produced by prey is directly related to the rapidity of pulsing [see also Myrberg et al., 1972].)

A diver often pulls a speared fish close to his body to minimize its struggles. This sounds dangerous, for the diver could get bitten accidentally if a shark tried to grab the fish under these circum-stances. But my observations and those of Gene Helfman (pers. comm.) tend to support the Palauan contention that this strategy often substantially lessens aggressiveness in grey sharks. It is less likely to be followed in the presence of lemon sharks, however, because according to divers, they are less likely to be dissuaded in this manner. If the diver is at the surface he will sometimes hold the speared fish above the water so that its struggles are less audible.

In other parts of the Pacific small blacktip sharks, *Carcharhinus melanopterus (matukeyoll)*, are often thought to be harmless. But they are treated with considerable respect by Palauans, and with good reason. (Raroians also fear this species [Harry, 1953].) There have been at least five documented attacks by this species in Palau (Randall and Helfman, 1973). None was fatal, but some were crippling. When a Palauan sees a blacktip heading in his direction he "barks" (this is the Palauan phrasing)—"wu' wu' wu' "—or stomps on the bottom of his boat, and the sounds seem to repel the shark. Reef blacktips have the worst reputations for stealing the divers' fish at night in shallow water. Reef whitetips, *Triaenodon obesus (ulup-suchl)*, represent little direct threat to divers unless speared, but will occasionally rip a fish off the diver's spear.

Spearfishing at Night

Many fish that are active during the day (such as parrotfish, surgeonfish, and wrasse) are inactive at night, making very easy targets as they rest on the bottom or in holes and crevices in the reef. The night diver thus usually spears fish at close range and in rough terrain where a gun's range and power are much less important than its maneuverability. Accordingly, small guns, three and one-half to four and one-half feet long, are generally used. The spear shaft is seldom attached by a line to the gun because fish speared at close quarters under these circumstances seldom have an opportunity to escape; the spear does not fully leave the gun and pins the fish to the bottom.

With only a flashlight for illumination the diver would easily lose the gun if it were to float when he released it to string his fish. So night spearguns are usually made of dense wood, such as ironwood, so that they are negatively buoyant. Similarly small but more powerful guns are used when divers specifically seek the large, powerful bumphead parrotfish as it sleeps. The spear for this gun is unbarbed to facilitate its rapid removal from the tough head.

A few more adventuresome underwater spearfishermen will search for fish outside the reef in fifteen to thirty feet of water at night. This is usually done from a boat. With the aid of the boatman the diver tries to get the fish off his spear and into the boat quickly so as to reduce the chances of attracting sharks. Large sharks are more abundant in shallow water at night than during the day. Divers therefore often find it necessary to move in the boat to a different spot every few minutes.

If a light is shined into the eyes of a shark and then moved smoothly away, the shark will frequently follow the beam, often at considerable speed. The newcomer to Palau is often skeptical of stories he hears about Palauan divers intentionally driving sharks directly into fellow divers using a beam of light. The stories are true, but the practice is not quite as reckless as it might sound and is not meant maliciously. Only small sharks (three to four feet) are used. And, as Ngiraklang states, "You only do this to a friend—as a joke. A stranger would get very angry." Larger sharks are sometimes disposed of at night by leading them at high speed in a beam of light toward the reef where they crash blindly into corals, discouraging them from further investigation of the divers if not killing them outright. For obvious reasons one does not leave one's light shining on another diver.

Estuarine crocodiles, *Crocodylus porosus (ius)*, are not commonly encountered by divers; Palauans shoot them with rifles on sight and they are not abundant. But they can be exceptionally aggressive at night, and divers say they fear them far more than any shark. One Palauan was attacked and eaten by a crocodile while night-diving in 1965 (Brower, 1974).

Hand Spears

Uses to which fishermen put the Palauan hand spear are changing. Until recently, as described earlier, fleet-footed younger fishermen routinely chased and speared rabbitfish and other fishes in shallow water on the reef flat. "But today," said Ngiraklang, "you can't get enough fish for a meal that way." At night Palauans also used to spear fish by torchlight from bamboo rafts. One of their favorite targets was the bumphead parrotfish, *Bolbometopon muricatus (kemedukl)*. It came regularly into shallow water to sleep facing into the current.

Spearing these fish by torchlight from rafts was not very efficient. Visibility was poor, many fish went undetected by sleeping in crevices or beneath underhangs, and the depth to which they could be exploited was very limited. But these problems were greatly reduced with the introduction of spearguns and high-powered underwater flashlights. Fishing pressure on bumpheads increased greatly as a consequence. Now most remaining bumpheads sleep in deeper water where they are out of range of raft fishermen and less accessible to divers.

Similarly other once-plentiful species that rested in shallow water at night have been either depleted or shifted their sleeping quarters to deeper water, according to fishermen.[11] The same problem has been reported from Tahiti (Bagnis et al., 1972), the Tuamotus (Ottino and Plessis, 1972), Wallis Island (Hind, 1969), and Ponape (McCoy, pers. comm.).

11. Changes in behavioral responses of fishes over time to new fishing techniques deserves more study than it has received. Decreased catches eventually experienced in a developing fishery are customarily treated by biologists solely as a manifestation of decreasing abundance of fish. Two other factors that may also be operating are learned avoidance of fishing gear and genetic selection for individuals less prone to being caught. Underwater spearfishing puts fishermen in a better position to observe the responses of their prey than any other type of fishing. And tropical spearfishermen around the world have observed that reef fish become markedly more wary in a matter of weeks in an area newly exposed to spearfishing. For example, groupers in unfished waters will commonly approach within two or three feet of divers, even accepting food from their hands. On heavily spearfished reefs, in contrast, they usually disappear in a flash into holes in the reef well before the diver gets within spear range.

Very heavy spears were also once used from canoes to spear dugong (*mesekiu*), which abounded and were a staple food sixty years ago (Kramer, 1929). Today they are an endangered species, depleted as a consequence of the ease with which they can be chased and speared in an outboard motor boat. I saw only one during my stay.

The introduction of outboard motors has created new employments for the hand spear. Although spearing fish from motorboats is not economical—gasoline is too dear—Palauans love high powered boats and spear from them for sport. (Nowhere else in Micronesia is gasoline burned so freely for the sake of speed. Ngiraklang, at 79, took my breath away more than once with his *kamikaze* approach to boating. One day I crashed clear through the seat of his boat as a result of his full-throttle assault on some rough water.) A by-product of these activities are several interesting insights into the behavior of marine animals that can, with the advent of fast boats, be chased down and speared.

A modern Palauan spearman wears polarized sunglasses and searches for fish from the bow of his runabout. In one hand he holds the taut bow line to steady himself. His other hand holds a spear, tip down, shaft resting on his shoulder. His success depends not only on his accuracy but also on the maneuverability of the boat and the responsiveness of the driver. The spear tip is lifted to signal to the driver that fish have been sighted. The shifting course the boat must take in the chase is telegraphed to the driver by pointing the spear. A number of fish, such as large wrasse, parrotfish, or snappers, sometimes "hole up" when pursued, taking refuge within a coral patch. The driver circles several times to discourage the fish from leaving its refuge. The spearman then dons mask and snorkle and moves in on the fish underwater. The head of the fish offers the best target. Palauans have found that hiding fish tend to rest with their heads facing into the current. The diver approaches the hole with this in mind.

In pursuing turtles speed has replaced stealth. The canoe fisherman paddled slowly and quietly up to a turtle hoping to get close enough to throw his spear before being heard or seen. Today's spearfishermen approach turtles at top speed. Hawksbills and small green turtles tend to head straight for deep water. Consequently they are approached, if possible, from deeper water so that they will have to

---

Divers in Palau and elsewhere swear that fish will allow unarmed divers to approach closer than divers carrying a spear. That is my impression too. Whether species that sleep habitually in shallow water "learn" to sleep in deeper water soon after being exposed to heavy night-time underwater spearfishing or whether their altered behavior is entirely the result of natural selection is not obvious.

run toward the boat. Hawksbills seem to have less stamina than greens and tend to give up quickly, making them comparatively easy targets. Larger green turtles often run a short distance, then circle the boat, apparently trying to confuse the pursuers. Eventually they either come up for air or seek shelter, in either case rendering them easy targets.

Some fishermen maintain that, more often than not, fish will run toward the sun when chased across a sand flat. When this happens they are hard to see in the glare and the driver tries to maneuver so as to put the boat between sun and prey.

Large barracudas are not chased by the prudent; several fishermen have had expensive damage inflicted on their boats by the teeth of a large barracuda that turned and charged its pursuers.

Milkfish, *Chanos chanos (mesekelat)*, "the strongest fish in Palau," provide challenging sport. When a school is sighted feeding on a shallow sand flat, it flees, taking a swift erratic course. Schooling functions to confuse predators, including man, by making it difficult to focus easily on a single target. Visually isolating an individual fish for long in a fast-moving school of milkfish is virtually impossible. And the chance of hitting a fish by throwing blindly into the school is very poor. So the hunters set out to separate a single fish from the school.

As shouting fishermen and alarmed fish careen across the reef flat the school eventually splits up. The boat follows the smaller group. This group in turn may split. Once again the pursuers chase the smaller group. Sooner or later a single fish will take off alone. The pursuers have now isolated their target.

After several minutes the lone fish slows down abruptly. Simultaneously it changes color, from silver to gunmetal blue. This appears to be a manifestation of stress; it is also seen in sick or roughly handled milkfish in the laboratory (Nash, pers. comm.). When this happens the spearman gets ready for his throw. But occasionally, before he can release the spear, the path of the fleeing fish and that of a school of milkfish will intersect. In this case the fish joins the school and accelerates to match the speed of its startled brethren. But its fate is generally sealed now, for it cannot hide among the school. Its much darker color marks it unmistakably and the hunters pursue it without difficulty.

Soon it slows down once again. Eventually it will turn slightly sideways for a moment, thereby presenting a larger target. At this instant the spearfisherman raises his spear and drops the bow line so that he can extend his other hand to sight the spear along it. His feet are wide apart bracing him against the gunwales. His knees are slightly bent. He does not lead the fish with his throw as a shooter leads a bird in flight; unlike the bird hunter he is travelling in the same direction and at the same speed as his quarry. He throws slightly behind his target to compensate for light refraction. He does not lunge forward for this would cause him to risk losing his precarious balance. Rather he twists his torso as he throws so as to impart extra momentum to his spear.

A second spear lies across his foot. If his first spear misses he flips up the second one and in an instant is ready for another throw. But Palauans possess an accuracy when throwing—graduating with age from marbles to rocks to spears—which I suspect (at the risk of annoying those who believe that all peoples are born not only equal but also identical) is a genetic gift. In short, although the target is small and moving and the boat is pitching, a second throw is often unnecessary.

In the Aimelik–Airai area of southern Babeldaob rays are particularly abundant on the reef flats and in mangrove channels. Hunting them on foot in shallow water with hand spears has been the tradition of one Palauan family for generations, and the method has not changed. Rays make depressions in the bottom when foraging for the crustaceans and molluscs they eat. A fisherman can often identify a ray by the characteristics of the depression it leaves behind. The fantail stingray, *Dasyatus sephen (ksous)*, creates an almost round hole similar in shape to its body outline. *Kutalchelb-eab*, an unidentified species, make an almost triangular hold similar to its outline. *Tebukbuk*, another unidentified species, leaves a deeper hole than the other two. If a feeding depression is made in a muddy or sandy bottom the edges collapse and the hole loses its identifying shape quickly. But when it is made in a bottom consisting of a mixture of mud and sand the hole tends to fill in only gradually, even in the presence of a strong current. A fisherman searching over this kind of bottom can therefore tell roughly how old a depression is and therefore how likely it is that the ray that made it will still be in the vicinity.

If he sees that a depression has been made recently he sets off in search of its maker. When rays take leave of a depression they generally do so in the same general direction they were facing when they made it. That direction is indicated by the shape of the depression; the deepest part of it is eccentrically placed at the point where the mouth of the ray was when making it. The spearman takes a zigzag path in the direction thus indicated.

If, when he encounters the ray, it is feeding, the first sign of it will be a cloud of suspended sediment. (This can be misleading, however, because sweetlips, *Plectorhynchus obscurus [bikl]* make similar feeding disturbances in the same general area.) If it is a *ksous* its tail will be seen sticking almost straight up through the sediment cloud. At other times, particularly around low tide in the morning, the ray will be found resting buried in the sediment with only tail, eyes, and gill holes showing.

There is an unresolved argument among ray fishermen in Palau and elsewhere as to whether one can subdue a ray faster by spearing it in one wing or between the eyes.

Cast Netting

The cast net, or *bidekill*, was introduced into Micronesia by Spaniards in the nineteenth century. It is a small circular net with light weights around the edge. The fisherman walks with the net hanging over one shoulder. The net is thrown so as to spread out horizontally and fall over a school of fish, the sinkers trapping them underneath it. Some cast nets have a drawstring around the edge. As the net is retrieved the drawstring gathers in the edge of the net beneath the fish so that the net assumes the shape of an inverted drawstring purse with the fish inside. Where the bottom is rough and jagged the net cannot be closed in this manner and cast nets without drawstrings are commonly used. In this case the fisherman gathers the fish from under the net one by one while the net remains spread out on the bottom.

Although cast netting looks easy, learning to throw it is just the beginning of learning how to use it effectively. A good *bidekill* fisherman can not only spot fish where the neophyte sees nothing but he can also often tell what kind of fish they are and thus how best to stalk them. The surface wake created by an invisible school of fish swimming near the surface, for example, provides clues to the identity of the fish that make it. Mullet swim rapidly with quick changes in direction and speed when they are not feeding. A unicorn fish swims deliberately and with greater independence of its school. When mullet are feeding in shallow water, head down, their silvery tails often break the surface. A number of other surgeonfish, rabbitfish, rudderfish, and emperors also advertise their presence when feeding in shallow water as their tails break the surface.

A fisherman stalks such schools by crouching low while pointing his feet and lifting them straight out of the water as he walks so as to minimize the noise he makes. Because the net is quite heavy when wet, this is a particularly physically demanding technique and is best done by well-muscled younger fishermen.

Some species, such as rudderfish, *Kyphosus* spp. (*komud*), feed beneath the crests of breaking waves. Under such conditions the fisherman moves during his approach only at the instant that a wave breaks, thereby hiding his movement from the fish. The net is also cast just as a wave breaks, obscuring the fisherman and the approaching net. Some fish feed in the foam just inshore of breaking waves. Here, once again, the throw is timed to take advantage of the momentary opacity of the water.

Knowing the paths that certain species habitually take as they move on and off the reef flat with the ebb and flow of the tides (see p. 49), the older, more experienced fisherman often stands and waits, much like a reef heron, for the fish to come to him. Some species are best caught with a throw net at low tide on the fringing reef flat because they tend to move into depressions. Others may be caught in deeper water and on any tide by casting from the reef edge, a cliff, or a dock.

The wind should be from the back or side of the thrower or the net will not spread well. The sun should not be behind the thrower, particularly if it is low in the sky, as the shadow of the fisherman and the approaching net will scare the fish. An overcast sky and light rain are the preferred daytime conditions for cast netting. The lack of shadows and the slightly disturbed surface of the water help conceal fisherman and net. Experienced cast net fishermen also fish at night, even in the absence of a moon. They know from experience where and when various species of fish will congregate, and they judge when to throw the net by listening for the sounds of fish breaking the surface.

Portable Fish Traps

Although Palauans once used at least thirteen different kinds of wooden fish traps (Kramer, 1929), only one basic type is used with much frequency today. It is usually three to four feet high and six to seven feet long. It may be made from sticks and vines or welded reinforcing bars and chicken wire, or a combination of both local and manufactured materials. In addition to a single entrance for fish at one end, there is a door in the top of the trap which is opened to remove the fish.

The trap is usually placed in shallow areas through which fish are known to migrate diurnally or with the tides. The traps are not pulled to the surface to be emptied, as is customary in many parts of the world. Rather the owner dives down, opens the door, and spears trapped fish one by one with a hand-held spear. The reason for this is that the trap is camouflaged by piling rocks around the sides and by creating a kind of rock arch in front of the mouth; if the trap were pulled to the surface it would take considerable time and effort to remove the rocks first and to replace them when the trap was reset.

Considering the long tradition of trap use in Palau it is surprising how divergent the opinions are among fishermen concerning

when, where, and how best to set them.[12] The general consensus
seems to be that the long axis of the trap should always be set
parallel to the prevailing currents. If it is set at an angle and the
currents are strong the free ends of the rods projecting inward in the
entrance funnel vibrate and scare away fish. (Japanese fisheries
scientists have similarly noted that fish nets sometimes vibrate in a
current and thereby scare fish [Nomura, 1980].)

When setting a trap the fisherman always brushes the sand at
the entrance with his hand as his last act before leaving it. This
custom is found in other Pacific island groups (e.g., Buck, 1930). In
Palau its original purpose seems to have been forgotten, for each
fisherman has his own tentative explanation for why he does this:
"to brush away the human smell," "to make the entrance look
well traveled," "for good luck," and the like.

Traditional Palauan traps were not baited. Some are baited today
but there is no consensus among the fisherman as to whether bait
really helps catch fish. (Biologists have found that bait has little
effect on trap-catch rates on Caribbean reefs [High and Ellis 1973;
Munro et al, 1971].) Certain species of fish attract others of their
species into traps, according to fishermen. Similar observations have
been made in the Caribbean by Munro et al. (1971) and High and
Ellis (1973).

Fishing with Dynamite

During World War II the Japanese government put Palauans to
work fishing to feed the troops. Catching enough fish was difficult.
There were several times as many Japanese to feed as there were
Palauans. The Japanese offshore tuna fishery had been suspended
and most Palauan canoes and rafts had been destroyed for fear that
they might be used to collaborate with the enemy. Serious hunger
threatened, possibly for the first time in Palau's history. Dynamite,
provided by the Japanese, was used by fishermen as a short-term
solution.

A fish bomb is made by packing a beer can about two-thirds full
with powder. The fuse is made from match head shavings that are
tamped down firmly into a thin bamboo rod using a thin stick. The
fuse is stuck into the powder and the can is then sealed with a layer
of mangrove mud. The length of the fuse is adjusted so that the bomb
will explode at a depth appropriate for the school of fish being
sought. Over the years, according to Ngiraklang, at least eight men
have been killed by premature explosions when making or using fish
bombs. Others have lost extremities or eyes.

12. The fishermen of Lukunor Atoll, Truk district, appear to have had a more detailed
rationale for where and how to set their traps (Tolerton and Rauch, 1949).

Dynamite not only kills target fish, but also kills many juveniles and destroys the reef habitat on which fish depend for food and shelter. As a consequence of its wartime use, fish stocks declined drastically according to Ngiraklang and other old men who were among those who used dynamite. Its use was made illegal (with the blessings of most of these men) after the war. But abandoned ammunition caches and unexploded bombs provided dynamite for some less principled fishermen for many years. Palauans said that in the early 1970s there were still caches of wartime dynamite carefully hoarded for dynamite fishing. Other sources, such as the Public Works Department, are also now available and the practice continues, albeit less often than in the past. Public opinion is increasingly against dynamite fishing and so it is restricted mainly to the districts of Koror and Peleliu where numerous islands and bays enable fishermen to use it unobserved and unheard.

Another destructive and illegal form of fishing used widely in the tropics involves pouring bottles of hypochlorite bleach into coral heads and caves, thereby not only killing the fish therein but everything else in the vicinity. This method seems only rarely to be used in Palau today.

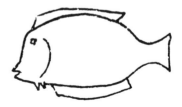

CHAPTER

RHYTHMS OF FISH AND FISHERMEN

3

Learning how to reckon moons and tides goes hand in hand with learning how to fish.

—*J. C. Cordell*

The average city dweller is seldom very conscious of the phases of the moon or aware of how closely they relate to many biological rhythms. When told of these relations he is liable to shrug them off as the products of superstitious minds. But many are real and no one is more aware of their reality than the traditional Palauan fisherman. The moon provides him with vital information concerning where, when, how, and for what to fish. Its phases accurately foretell not just the timing and approximate height of the tides, the strength and direction of the tidal currents, the brightness of the night, and the accessibility of different fishing areas, but also the locations, behavior, and vulnerability to capture of many species of fish.

Throughout Oceania anthropologists and early Western travellers often noticed that island fishermen caught more fish during some phases of the moon than others. In addition to numerous brief, published allusions to this phenomenon, I found published lists of the good and poor fishing days of the lunar month in Tahiti, Hawaii, and Raroia (table 1). Good fishing days cluster around new and full moons in all of these lists, whereas the intervening periods are generally times of relatively poor fishing.

The names given to certain days of the lunar month by other Pacific islanders add to this picture. On Namoluk Atoll, Caroline

Islands, for example, the night before the new moon is called *Otolol*, meaning "to swarm." Fish are said to swarm and be easier to catch at this time (Girschner, 1912). In the Gilbert Islands, similarly, the name of the day after the new moon means literally, "it swarms of fish" (Grimble, 1931).

There is a cryptic suggestion in one of these accounts concerning the reason for this lunar periodicity in fishing success. Handy (1932) stated, "If Tahitian teaching concerning the habits of fish, is trustworthy—and it would be strange if it were not, for the food supply depends on fishing more than anything else—then the run of fish in the Society Islands is governed by the phases of the moon." "Runs" of what kinds of fish? Where? And for what purpose?

Tagging studies carried out in various parts of the world have led many biologists to conclude that during their adult lives most reef fishes tend to remain in small, circumscribed areas (e.g., Ehrlich, 1975; Sale, 1978). How can these observations be reconciled with Handy's statement that fish "runs" are an important aspect of reef fisheries?

One of my objectives in Palau was to determine whether lunar fishing rhythms similar to those reported from elsewhere in Oceania existed and, if so, to investigate their causes. I soon found that such rhythms do occur (table 1), that many reef fish do indeed "run," and that the phenomenon is so important to Palauans that almost any village school child can describe it. The following incident serves to introduce the explanation.

In March 1976, one day before the new moon, I encountered hundreds of groupers of two different species massed in an area of roughly 1,000 square yards in and around the mouth of Ngerumekaol Channel which cuts part way through the fringing reef near Ulong Island. Because groupers are sought regularly by spearfishermen in Palau, they have become wary. Normally they retreat swiftly into holes in the reef as soon as they see an approaching diver. Larger individuals are particularly elusive. Yet on this occasion the fish were surprisingly docile; I was almost able to touch them. If I had used a speargun I could have filled the boat with them in an hour.

This was a lunar spawning aggregation—one of many familiar to Palauan fishermen. Ngiraklang alone told me the lunar spawning cycles of forty-five species. Only a few cases of fish with lunar spawning rhythms had been recorded in the scientific literature and biologists were unaware of their prevalence among coral reef fishes. Why were these rhythms so important in this particular environment? Gradually other distinctive aspects of the behavior of spawning reef fish emerged from fishermen's descriptions. Together with my own subsequent observations they suggested an answer.

TABLE 1.   Good Fishing Days of the Lunar Month in Oceania

Lunar Day	Raroia[a]	Tahiti[b,c]	Hawaii[d,e]	Palau
1 (new moon)	x	x x	x	x
2	x	x x	x	x
3	x	x x	x	x
4	x	x x	x	x
5	x	x x	x	x
6		x x	x	
7				
8				
9		x		
10				x
11			x x	x
12		x	x ?	x
13	x	x x	x x	x
14	x	x x	x x	x
15 (full moon)	x	x x	x x	x
16	x	x x	x x	x
17	x		x x	
18	x		x x	
19			x x	
20			x	
21				
22				
23				
24	x	x x	x	
25	x	x x	x x	
26	x	x x	x x	
27	x	x x	x x	
28	x	? x	x x	
29		?	x x	x
30		?	x x	x

[a]Danielsson, 1956.   [c]Stimson, 1928.   [e]Taylor, 1957.
[b]Handy, 1932.   [d]Handy et al., 1972.

A typical spawning run begins a day or two before spawning starts. Individuals of a species will often first converge at a predictable location. If these fish are species that live in the lagoon or on the reef flats they will then generally migrate seaward en masse, using routes well-known to fishermen. (Other Pacific island fishermen have also noticed this tendency of various reef fishes to migrate seaward in order to spawn [e.g., Akimichi, 1978; MacGregor, 1937].) Their destination is usually a particular area along the outer reef edge, over the outer reef slope, or in and around a channel cutting into the outer reef slope. Here they aggregate for several days, often joining other schools of the same species, and spawn.

The aggregation then breaks up and the fish return, individually or in small schools, to their points of origin. It is apparently because these fish return to their normal haunts shortly after spawning that tagging studies have generally failed to uncover their spawning migrations. Unless researchers managed to catch tagged fish during their short spawning runs they would be unaware that the fish ever left the area where they were tagged.

Typically, according to Palauans, reef fish spawn while the tide is going out.[1] Tahitian fishermen made the same observation. According to Nordhoff (1930), "The natives say, and I believe with reason, that nearly all of the fish that haunt the lagoons and reefs spawn at times when the currents sweep the eggs far out to sea. The result is that hundreds of different kinds of fry hatch out in the open Pacific."

1. So prevalent is the apparent influence of the moon and associated tides on the behavior of fishes that Palauans believe they also influence human behavior. And indeed they once apparently did in at least one way unrelated to fishing. Palauans were traditionally given to dancing and revelry on nights around the full moon—so much so in fact, German plantation managers passed a law requiring them to stay home for the three days around the full moon so that the coconut seedlings on their plantations received adequate care (Kramer, 1929).

Palauans also believe, even today, that people generally die on incoming tides. (In certain other parts of the world just the opposite belief is held [Beck, 1973; Dickens, *David Copperfield;* Harley, 1970].) They also believe that women, like reef fish, generally give birth on outgoing tides. So prevalent is this belief that a tide table is posted in the delivery room in Palau's hospital, and nurses are heard to comfort women in labor by telling them "the tide will turn soon and then the baby will come" (Dr. M. L. Jung, pers. comm.).

I checked two years of birth records from the hospital to find out whether this belief might have sufficient power to influence birth patterns. I found that equal numbers of births occurred during incoming and outgoing tides. However I also found that 25 percent more births occurred during neap tides (around half moon periods) than during spring tides (around new and full moons). Statistically the likelihood of this occurring purely by chance in either year was less than one in twenty. Günther (1938) found a similar though much less pronounced lunar rhythm in human births in Cologne, Germany. However a number of other investigators have reported a variety of contrasting results—different weak lunar rhythms or no lunar rhythms—in other localities (see reviews by Lieber 1978; McDonald, 1966). I found no lunar birth rhythm similar to that in Palau when I analyzed 1976 and 1977 birth statistics for Hawaii.

Why should a pronounced lunar birth rhythm occur in Palau when only relatively slight or no lunar periodicity has been demonstrated in other localities? Osley et al. (1973, p. 414) predicted that "greater lunar effects (on birth timing) may be expected in cultures without electricity, where the moon provides the principle nightime illumination and permits social mobility." Palau is only partially electrified, and village generators are usually turned off well before midnight. But if this were the synchronizing mechanism, peak birth periods in Palau should occur around full moon, not around half moon periods, because the human gestation period is precisely nine lunar months (Menaker and Menaker, 1959).

Perhaps the answer lies in the markedly collective nature of the activities of Palauan women; they typically carry out their daily house and garden routines in groups. McClintock (1971) and others have demonstrated that a synchronization of menstrual cycles occurs among women who live or work closely together. In addition, McClintock suggested that dark-light patterns, including those due to lunar rhythms, may function to lock the menstrual cycles of such women in phase.

A pattern had thus emerged from these accounts. Many reef fishes spawn at the outer edge of the reef, near deep water, on outgoing tides. And they spawn during those portions of the lunar month in which tidal ranges and tidal currents are greatest—around new and full moons. The locations and the timing of spawning thus combine to provide ideal circumstances for flushing the eggs off-shore and away from the reef habitat. But of what value is a reproductive strategy that favors the transport of eggs and the hatching of the young over deep water in areas removed from the adult habitat?

Coral reef communities abound with fishes, corals, and other animals that feed on zooplankton, including drifting fish eggs and larval fishes. When seawater flows across a shallow coral reef for a distance of a few hundred yards, reef predators remove 50 to 90 percent of the zooplankton (Glynn, 1973; Johannes and Gerber, 1974; Tranter and George, 1972). The eggs of most reef fishes hatch into planktonic larvae. The hatchlings generally live in the plankton for several days to several months before taking up residence in the reef community. For such tiny animals with limited swimming ability the reef community is a jungle of grasping tentacles and waiting mouths. It is a place where being eaten is not just a possibility; it is a probability. Both the timing and location of spawning of many reef fishes thus appear to have evolved to enable their eggs and larvae to escape the intense predation pressure characteristic of their adult habitats.

Many species—groupers, milkfish, mullet, rabbitfish, jacks—are unusually docile and approachable when in spawning aggregations. Palauans refer to these fish as being "stupid" at such times. Other Pacific island fishermen I interviewed were also familiar with this spawning stupor. However I can find no general discussion of it in the biological literature.[2]

As a consequence of this atypical docility, and of the unusually large numbers and high densities of fish converging at predictable times and locations in these spawning aggregations, they provide exceptional fishing. So fishermen flock to the spawning grounds with lines, nets, spears, and, sometimes, dynamite. Because most such species aggregate to spawn around new or full moons (see appendix 1) this accounts in large measure for the high catches at these times of the lunar month.

2. Some temperate zone fish also exhibit reproductive stupor (e.g., Reighard, 1920; Znamierowska-Prüffer, 1966, p. 91). The phenomenon was also known to the Greeks at least nineteen centuries ago; Pliny the Elder said of mullet, "at the time of coupling . . . their salacious propensities render them unguarded." See Helfrich and Allen (1975) for support for this statement.

The stomachs of some of these species are completely empty prior to spawning. (Cessation of feeding prior to spawning has also been observed in a variety of temperate zone fishes [e.g., Iles, 1974; Shul'man, 1974].) But they nonetheless often take baited hooks or lures just as nonfeeding salmon do on spawning runs (e.g., Boyd et al., 1898). And they begin to feed voraciously immediately after spawning, according to fishermen.

As described later in this chapter, other factors associated with new and full moons influence fishing success. But fishermen maintained, and my observations confirmed, that spawning aggregations were the major source of big catches around new and full moons. (Rabbitfishes are an important exception. Two species that are very important to fishermen in some parts of Palau spawn and are caught in particularly large numbers beginning several days after the new moon [see appendix A].) There is thus correspondence in Palau with geographer Cordell's (1974) suggestion that among artisanal fishermen in coastal Brazil a lunar-tide system of determining fishing strategy has been perfected as a consequence of spawning periodicity.

Learning and committing to memory the timing and location of these aggregations is an essential part of becoming a good fisherman in Palau. And many of the older fishermen retain the formidable memories characteristic of preliterate peoples, despite their exposure to written languages since Japanese times. As a consequence some of them are remarkable repositories of knowledge about lunar spawning aggregations.

By examining the gonads of various species through the lunar month and by visiting the sites of reported spawning aggregations I found that the information provided by Palauan fishermen was highly reliable. Among those species that I investigated only one, the milkfish, mesekelat (Chanos chanos), failed to aggregate with ripe gonads at the time and place predicted by the fishermen—a puzzle that is described in appendix A along with a species-by-species account of other reported lunar spawning rhythms and aggregations.

Although lunar reproductive rhythms are apparently rare among strictly terrestrial animals, some land crabs return to the sea on a pronounced lunar cycle to release their eggs. The best known example of this among Palauans is Cardisoma hirtipes (rekung el beab). This fist-sized crab lives in holes in the forest floor, coming out at night to feed. This routine is interrupted beginning several days before the full moon, mainly during the months of the southwest monsoon. Females with eggs abandon their holes and set out on a mass migration toward the sea. At this time women and children collect sacks full of them with ease as they cross the roads of Angaur and Peleliu. At dusk, starting about two days before the full moon, the crabs emerge from the edge of the forest and move warily down sand beaches to the edge of the water.

I watched this phenomenon in June 1977 as the full moon rose on a quiet night off Honeymoon Beach on Peleliu. Hundreds of crabs were moving quietly out of the underbrush and down the sand. My approaching movements, though careful, sent them scurrying back to shelter. As I settled down they once again began this strange march toward their original home. When a crab reached the water's edge it walked into the oncoming waves. The instant it touched the water it began to act as if oblivious to human activity, even seeming to ignore the bright light I shone on it. The first incoming wave generally knocked it over. Quickly righting itself as the wave receded, it dug its rear walking legs into the sand, braced itself for the next wave, and elevated the front end of its body. As the next wave washed over it the crab flapped its abdomen and shook its pincers vigorously up and down in unison, actions that shook loose the larvae that had just hatched from the embryos carried beneath its abdomen. The larvae were all released in a dark stream during the advance and retreat of two waves. The crab then moved back onto the beach, having spent just a few seconds in the water. It immediately regained its customary wariness and moved up the beach and back into the forest.

The temporary loss of caution by land crabs while releasing their larvae is similar to the stupor of spawning reef fishes. (One wonders whether this phenomenon has contributed to the widespread belief in "moon madness.") It is also characteristic of a number of other species of land crabs in Palau (Gene Helfman, pers. comm.). In such situations the animals are very vulnerable to predation—the crabs far from their normal shelter, the fish packed densely on spawning grounds and often besieged by sharks and other predators (e.g., Helfrich and Allen, 1975). If either fish or crab fled from predators at this time reproduction would be disrupted. Reproductive stupor may thus be a "flight override" mechanism that ensures that the aggregations will complete their reproductive act successfully even though some individuals will be lost to predators. Sea turtles, though threatened when mating by fishermen and sharks, and when laying eggs by turtle hunters, also seem oblivious to predators at these times (e.g., McCoy 1974), perhaps for the same reason. Reproductive stupor does not appear to occur among smaller reef fish species that spawn in small groups and stay close to shelter to which they retreat quickly if approached (Johannes, 1978b).

The lunar rhythm of larval release in the land crab, *Cardisoma hirtipes*, was first described in the biological literature in 1971, for the Ryukyu Islands south of Japan (Shokita, 1971). But it is clear from accounts of travellers and anthropologists that the phenomenon is widespread and well known among the peoples of Oceania and was recorded by Westerners more than ninety years ago. Land crabs have been reported migrating to the water's edge on a lunar cycle (usually around the full moon) in Samoa (Pritchard, 1866), Fiji (Woodworth, 1903, 1907), Guam (Amesbury, pers. comm.; Safford, 1905), New Britain (Brown, 1910), Kosrae (Kusaie [Sarfert, 1919]), the Solomon Islands (Ivens, 1927), Tahiti (Stimson, 1928), the Cook Islands (Beaglehole and Beaglehole, 1938), Ponape (Bascom, 1946), Ifaluk (Bates and Abbott, 1958), and Satawal (McCoy, pers. comm.). The same phenomenon also occurs on Tobi, Yap, and Truk according to fishermen I interviewed on those islands. On Lamotrek Atoll the traditional name of the second day after the full moon means "shoo," referring to the belief that the moon shoos the land crabs back into their holes after several days spent on and near the beach (Clifton, pers. comm.). According to Trukese fishermen the traditional Trukese name for the night of the full moon is *bonung aro*, meaning "night of laying eggs," referring specifically to the lunar rhythm of the land crab.

By releasing their larvae at the peak of the highest spring tides land crabs maximize the offshore flushing and dispersal of their larvae, thereby reducing their loss to the numerous predators who live in shallow reef waters. Releasing the larvae at night when visibility is low further reduces their vulnerability to predators. Similarly green and hawksbill turtles lay their eggs at night. The eggs are covered and hidden from predators before daybreak. Many species of reef fish also spawn at or after dusk according to Palauan fishermen; often fish caught late in the afternoon from a spawning aggregation will be ripe, whereas those caught early the next morning are found to be spent.

But why are full moon spring tides apparently more suitable than new moon spring tides for these reproductive migrations? Certainly the migrating crabs and their larvae would be less visible to predators during the dark nights around the time of the new moon. The answer may lie in the need for some kind of compass during the crabs' reproductive migrations, particularly if they live some distance from the water. On Peleliu some crabs appear to move as much as a kilometer from their burrows through dense underbrush on reproductive migrations and are quite selective about the locations along the beaches at which they release their larvae. In some localities this species may live several kilometers from the sea (Minei, 1966).

Some decapod crustaceans can use the moon as a light compass (see review by Creutzberg, 1975). Schöne's (1963) work suggests that the crab, *Talitrus saltator*, orients specifically to polarized moonlight. Waterman and Horch (1966) discovered that a Florida member of the same genus as Palau's *rekung el beab* can detect polarized light. This species (later found to be *Cardisoma guanhumi*; T. Waterman, pers. comm.) also makes reproductive migrations that peak around the time of the full moon (Gifford, 1962).

These observations are all consistent with the possibility that *Cardisoma hirtipes* release their larvae around the time of the full moon, rather than the new moon, because moonlight (or perhaps specifically polarized moonlight) provides a celestial compass that guides them to and from the particular beaches on which they shed their larvae. The moon is not only brightest around the time of the full moon, but, in contrast to other portions of the lunar cycle, is also visible throughout the night.

Seasonal Rhythms

Not far from the equator, and bathed throughout the year by waters of almost uniform temperature, Palau experiences little of the annual weather cycles we associate with the seasons in the temperate zone. The direction of the prevailing winds changes with the seasons, but temperature, humidity, rainfall, and day length are relatively constant. The physical environment thus provides only subtle clues to the seasons. Nevertheless it is clear that life responds to these cues. Seasonal reproductive cycles are everywhere and obvious. Trees blossom[3] and birds nest according to regular seasonal rhythms well-known to Palauan villagers (e.g., Klee, 1972), and every season brings predictable changes around the reefs and lagoons.

Seasonal environmental rhythms are charted traditionally in Palau, as in the rest of Oceania, using a calendar based on the lunar month.[4] Although lunar spawners lay their eggs during the same portion of the lunar month year after year, their spawning dates wander all over the Gregorian month. For example, the *kotikw*, *Gerres oblongus*, begins to lay its eggs every year on the new moon of the first month of *ongos* (Palauan New Year's Day). But if we date the event according to the Gregorian calendar it lays them some-where in October or November with a twenty-nine-day range from year to year. In the context of fishing in Palau it can thus be seen that a lunar calendar is more useful than the Gregorian calendar. The older men are familiar with the Gregorian calendar. But they use it mainly for such things as keeping appointments with younger

---

3. Seasonal variation in plant growth is not limited to the land. Kanda (1944) lists nine species of benthic marine algae that exhibit distinct seasonal periodicity in Palauan waters. Ngiraklang noted that seagrasses near Ngeremlengui bloomed mainly in March and April. In April of both 1976 and 1977 I observed several square miles of floating tufts of *Hormothamnion* sp., a benthic, nitrogen-fixing blue-green algae of the shallow reef, along the southeast coast of Babeldaob.

4. Hidikata (1942) first described Palau's traditional calendar. Klee (1972) has redescribed the Palauan system of time reckoning in considerable detail.

relatives, very few of whom have learned to use their own traditional calendar.

The *ongos* months begin in late October or November, on the day of the new moon two weeks after the only full moon of the year which, for a few minutes at its zenith, completely obscures the constellation Pleiades.[5] On the first nights of *ongos* the moon appears as an arc tipped toward the north and tipped further in this direction than in any other month of the year.

The traditional Palauan calendar has twelve lunar months. Because a lunar month averages twenty-nine and one-half days, twelve lunar months add up to 354 days—about eleven days short of the solar year. Thus, if each lunar year had only twelve months, the lunar calendar would shift around the seasons; in each successive year a given lunar month would begin eleven days earlier relative to the seasons and to the Gregorian calendar. A lunar year that began in early November in 1980 would commence in March by 1990. The first day of the new year would slide through the entire cycle of seasons in about thirty-three years. Because a major purpose of most lunar calendars was to help their users keep in tune with environmental rhythms, including those of the seasons, some adjustment of the twelve-month lunar year was obviously necessary.

This need was traditionally met in Oceania by inserting a thirteenth lunar month in the year about once every three years. Generally this insertion was accomplished unconsciously (Nilsson, 1920). This is the case today among the few remaining old Palauan fishermen who still use the traditional Palauan calendar. Unaware of the need for the periodic insertion of an extra month in their calendar these old men nevertheless unwittingly do so at the appropriate time.[6] To them the new year simply starts only when the stars and moon are "right," as described above, no matter whether twelve or thirteen lunar months have passed since the previous year began. When the beginning of the thirteenth lunar month does not coincide astronomically with the new year—that is, when the new moon is not properly "tipped" and the Pleiades are not lined up properly with the moon—the old men decide they must have lost count. They suppose that what they had believed to be (and was) the twelfth lunar month, was, due to an error in their counting,

5. Klee (1972) states mistakenly that the moon obscures the little dipper, *Ursa minor,* at this time. But, as seen from Palau, the little dipper remains low in the northern sky at all times and the moon in its transit from east to west never comes near it. My interpreter in Ngeremlengui mistakenly translated the Palauan word, *Mesikt,* as "little dipper;" this was probably also the source of Klee's error. Kramer (1929) identified *Mesikt* as the Pleiades.

6. Hidikata (1942) was unable to determine how Palauans made this correction. Curiously Klee (1972) does not allude to the problem.

only the eleventh month. Thus about every three years they insert
an extra "twelfth" month to make up for their "mistake."

When I lived in Ngeremlengui, the fishing was so poor for the
last few months of the Palauan year, due to strong onshore winds,
that the fishermen lost track of time and simply awaited the
appropriate signs in the heavens to proclaim the beginning of the
new year. To a modern reader with calendars in kitchen, office, and
wallet this may seem surprisingly absent-minded. But the traditional Palauan uses the lunar calendar mainly to keep track of the
behavior of the fish he seeks; when fishing is bad there is little
reason for keeping exact track of the lunar months.

A similarly unconscious calendar correction is made in connection with the number of days in the lunar month. In some months
thirty days elapse between new moons; in alternate months twenty-
nine days elapse. But old Palauans believe that all lunar months have
thirty days. The new month begins, however, like the new year—
when the moon is "right"—regardless of how many days have
elapsed since the last new moon.

During *ongos* months (roughly October through April) the
waters are relatively calm on the fishing grounds of Ngeremlengui
because they are in the lee of the prevailing easterly winds. Consequently most types of fishing are more productive during these
months. Fishing success is further enhanced because more fish
spawn during these months. During most *ngebard* months (roughly
May through September, the season of westerly winds) the prevailing winds blow onshore and the waters off Ngeremlengui are rough
and murky and fishing is generally poor.

The waves crashing on the outer reef edge during the rougher
portions of the *ngebard* months make the floors of the houses in
Ngeremlengui, three miles away, vibrate sensibly. The roar can be
heard on the opposite side of Babeldaob, nine miles away on the far
side of hills several hundred feet high. When particularly high swells
are coming in on a high tide their energy is not all exhausted against
the outer reef. The surge they create carries into shallow water
creating tide-like oscillations, with periods of a minute or two,
extending far up the tidal creek at Ngeremlengui. When a fisherman
sees the creek waters behaving in this fashion, he need not walk
through the mangroves down to the edge of the reef and look out to
sea to know that the day might better be spent on mending nets and
making traps than on fishing.

On the east side of Babeldaob the good fishing seasons are
reversed. Fishing is poor during the *ongos* months because of
onshore winds. Fishermen working out of Kayangel, the northern

and southern tips of Babeldaob, Koror, Peleliu, and Angaur are less restricted by rough weather. They can generally find sheltered waters in which to fish whichever way the wind blows.

Prevailing winds and currents bring significantly more drifting logs past Palau during the *ngebard* season of easterly winds. Because sharks tend to concentrate around these logs (see chapter 7) this was the season for offshore shark fishing, *oungeuaol*, off the east coast of Babeldaob. Such fishing was not carried out off the west coast of Babeldaob, probably because on that side of the island the season for logs coincided with the season of strong onshore winds and rough water.

Although spawning of some reef fishes occurs throughout the year in Palau, unusually large numbers of species, and of individuals within species, reproduce during the spring. According to Ngirak-lang and other fishermen a noticeable increase in spawning activity occurs in *Tmur ra Ongos*, the first month of the Palauan year, which begins in October or November depending, as discussed earlier, on the relation of the moon and the Pleiades. Spawning activity is fairly high through January and February, then increases to an annual peak in March and April. From June through September spawning activity of most species of food fishes (groupers are a notable exception) is comparatively low (see appendix A).

I wondered why there should be a strong multispecies spawning peak at a period of the year not marked in Palau by obvious changes in environmental factors that are often associated with spawning in temperate waters. My curiosity increased when fishermen from Yap, Truk, Ponape, Nukuoro, the Mortlocks, Tobi, the northern Marianas, and Majuro—Micronesian islands separated by as much as 2,500 miles—all subsequently told me the same story: around their islands more fish spawned between February and May than at any other time of year. Later I found in the biological literature that similar seasonal spawning peaks were widespread in the coastal tropics, being recorded from as far away as Malagasy (Madagascar) in the Indian Ocean and Jamaica in the Caribbean, but that none of the writers who had described such a peak had noted that it was more than just a local phenomenon.

Spring spawning peaks are well known in the temperate zone where rising temperatures and high plankton productivity provide favorable conditions for larval survival. But such is not the general case in the tropical areas where spring spawning peaks have been found. Here, collectively, there is no correlation between temperature or plankton productivity and "springtime." Nor was there a correlation between these spawning peaks and rainfall (high rainfall triggers spawning in many freshwater fishes). What was the explanation for this widespread tropical phenomenon that Palauan fishermen had led me to uncover?

I looked for correlations between these spawning peaks and other environmental variables. Ultimately I found that spring (or, in some cases, spring and fall) reef fish spawning peaks occurred at times of the year when prevailing winds and/or the prevailing currents were weakest. They coincided with times of the reversing of monsoon winds, the shifting of tradewinds, or with changes in the direction of wind-driven prevailing currents that are linked in turn to changes in trade wind or monsoon wind patterns. In Palau and other islands in southern Micronesia, for example, the spring spawning peaks coincide with an annual weakening of the prevailing ocean currents that bathe the islands (Wyrtki, 1974). In two areas (the Great Barrier Reef and Madras, India) spawning peaks occurred at times of the year other than spring or fall. Here typical monsoon or trade wind patterns do not occur, but reduced wind velocities *do* occur—at the same times of the year as the spawning peaks.

My next question was, "What do reduced prevailing ocean currents and reduced prevailing winds have in common in the way they affect coastal waters where reef fishes spawn?" One obvious answer is that either one will result in the reduction of water transport near the coast. But of what value is this to fish?

The eggs of most coastal marine fishes in the tropics hatch into planktonic larvae that are more or less at the mercy of the currents for several days to several months. As discussed earlier, spawning and hatching often occur at times and places favoring the transport of eggs and larvae to deep water outside the reef. It appears that his pattern has evolved to enable the larvae to escape the intense predation pressure in coral reef communities.

When the larvae attain a size and mobility that increase their chances of avoiding reef predators significantly, they move back into a suitable shallow-water habitat, provided they can find one. The existence of larval reef fish many hundreds of miles from the nearest reef indicates that some do not. Larval and juvenile reef fishes are sometimes even swept into temperate waters during the summer in some areas. Such individuals are doomed whether or not they find shallow water, having been transported to areas that will become much too cold for them in the winter (see, for example, Cresswell, 1979; Dooley, 1972). Others may survive but be unable to reproduce (e.g., Allen et al., 1976; Hobson, 1972b).

Thus, although an offshore pelagic existence reduces the chances of larvae being eaten, it increases its chances of being unable to find an appropriate shallow-water habitat to colonize. Reef fish larvae that hatch at a time of the year when the movement of ocean waters near the coast is reduced therefore have a better chance of remaining close enough to a suitable reef habitat to be able to return to it when partly grown.

The idea that the spawning strategies of many reef fishes help ensure the retention of their pelagic larvae in the general vicinity of, albeit beyond, the reef is supported by additional information volunteered by Palauan fishermen. The best-known spawning ground in Palau for a wide variety of reef and lagoon fishes is in the vicinity of an underwater promontory off the southern tip of Peleliu, the southern-most island within the reef that encircles most of the archipelago. Prevailing currents in the area tend to set to the south, away from the archipelago. This, therefore, might appear to be an unlikely spot for reef fish with pelagic larvae to spawn; it would seem that their progeny would have minimal chances of returning to their natal reefs. Palauans report, however, that a circular current, or "gyre," exists off the reef there. The evidence for this, although unconventional, is rather convincing. During World War II intense fighting occurred on Peleliu between American and Japanese forces. During one skirmish much blood was spilled by troops storming the beach. Rather than disappearing, the red water moved south to the southern tip of Peleliu, formed a large, almost stationary patch and remained visible for several days. The beach is now known locally as "Bloody Beach."

Ngiraklang also volunteered that a number of species of reef fishes, particularly parrotfishes, typically spawn off other underwater promontories along Palau's outer reef edge. Munro (1974a), Randall and Randall (1963), and Colin and Clavijo (1978) have similarly noted the tendency of various reef fishes to spawn off promontories on the seaward edge of reef systems. The gyre near the tip of Peleliu is not just a local anomaly: gyres typically form off points of land or submarine promontories (e.g., Defant, 1961; Hamner and Hauri, 1977; Hattori, 1970; Laevastu, 1962). Such promontories thus appear to be unusually favorable locations for increasing the probability that oceanic reef fish larvae will ultimately be returned to the reef periodically on local currents. Sale (1970) and Emery (1972) have pointed out the function of nearshore gyres in facilitating the return to shallow water of the oceanic larvae of coastal forms.

Palauan fishermen's accounts of spawning rhythms and behavior had thus led, via a trail of interviews in other parts of Micronesia and subsequent library work, to the discovery of some fundamental and distinctive aspects of reproductive strategy common to many tropical marine fishes. This subject is discussed in more detail elsewhere (Johannes, 1978a).

Calendar-Correcting by Marine Animals

We have seen that the sun directs the choice of spawning months of many reef fish whereas the moon directs the choice of spawning days. It is essential, therefore, that fish, like fishermen, have some means of reconciling their lunar and solar timing systems. Just what physiological mechanisms are involved is a mystery. But there is no doubt that they do it—like fishermen—by perodically inserting an extra lunar month into their internal reproductive "calendars."

Different species do not make this adjustment in synchrony with one another nor with fishermen. Consequently, as Palauans acknowledge, fishermen are occasionally one month early in predicting the commencement of spawning of a species. Most species of food fish in Palau spawn several times per year over a period of several months. Because fishermen's predictions are uncertain only for the month in which spawning commences, the practical consequences are not as troublesome as they would be if the fish spawned only once a year.

To examine this phenomenon of animal intercalation in any detail we require a long sequence of records of the annual dates of the commencement of spawning of a species. No such records appear to exist for reef fish. But records of the spawning dates of the palolo worm, a coral reef delicacy exhibiting pronounced lunar reproductive periodicity in Samoa and Fiji, serve to illustrate the kind of problems that the fishermen may be expected to encounter in predicting the commencement of spawning of a species exhibiting both seasonal and lunar reproductive rhythms.

Reproductive swarms of palolo occur for two days during the last quarter of the lunar month, in October and November in Samoa and in November and December in Fiji. These swarms occur at intervals of twelve lunar months in roughly two out of every three years, and at intervals of thirteen lunar months in the other years (table 2). The sequence of twelve- and thirteen-month intervals is not regular however. One cannot predict with certainty when the "usual" twelve-month interval will be replaced by a thirteen-month interval. Islanders were unaware of this complication and simply assumed that the annual palolo swarms would begin every twelve months. It is almost certainly for this reason, that, as a Samoan told Whitmee (1875), "mistakes have been made by the Samoans in predicting the moon [i.e., lunar month] in which Palolo will appear.

TABLE 2.   Intervals, In Lunar Months, Between Annual Commencements of Spawning of the Palolo Worm

	Samoa[a]		Fiji[b]
Year	Synodic months since Previous Spawning	Year	Synodic months since Previous Spawning
1943	12	1898	12
1944	12	1899	13
1945	13	1900	12
1946	12	1901	13
1947	12	1902	13
1948	13	1903	12
1949	13	1904	12
1950	12	1905	13
1951	12	1906	12
1952	13	1907	12
1953	12	1908	12
1954	12	1909	13
1955	12	1910	12
1956	13	1911	12
1957	13	1912	13
1958	12	1913	13
1959	12	1914	12
1960	12	1915	12
		1916	12
		1917	12
		1918	13
		1919	12
		1920	13
		1921	12

[a]Data from Caspers, 1961.
[b]Data from Corney, 1922, cited in Korringa, 1947.

We usually know the day, but are often in error as to the moon and expect it too early."

To make matters even more complicated, the ratio of two twelve-month lunar years to one thirteen-month lunar year does not reconcile lunar and solar calendars perfectly. An additional thirteen-month lunar year must be inserted every twenty-eight solar years. Reproductive records have not been kept for long enough, even for the palolo, in order to demonstrate the existence of this secondary correction in spawning rhythm. But it almost certainly exists.

Otherwise the primary correction described above would serve no purpose. Each year spawning seasons would begin about one day earlier in terms of solar time. A spawning season that began in October 170 years ago would now commence in March.

A similarly complex scheme of "animal intercalation" can be anticipated among fishes with lunar reproductive cycles. Because many reef fishes have life spans of only a few years (e.g., Munro, 1974a), this twenty-eight-year calendar correction would not be employed in every generation. For example, in a species with an average life span of four years and a single spawning season per generation, the correction would be made only once per seven generations. Discovering the physiological processes involved here would be a fascinating undertaking for biologists.[7]

Seasonal Circumnavigatory Migrations

At predictable times of the year certain species of fish move past Ngeremlengui in large numbers traveling north. Shortly thereafter villagers on the other side of Babeldaob see schools of the same species traveling south. Milkfish and certain species of snapper, barracuda, herring, mullet, and unicornfish are involved in such movements at different times of year. The regular sequences in which sightings of such schools occur near different villages indicate to fishermen that they must all be seeing the same fish and that they are traveling clockwise around Babeldaob and the Rock Islands.

Milkfish, *mesekelat,* reportedly migrate north from their nursery area at Peleliu along the west side of Palau to feeding grounds on the sand flats north of Babeldaob. They subsequently return south along the east side of Palau to spawn at Peleliu. The purpose of the migrations of the other species is not clear. (Some additional details concerning these migrations are given in appendix A.)

Other tropical island fishermen in both Pacific and Atlantic waters have reported similar migrations of certain fishes (e.g., Diaz,

---

7. The existence of genes that convey instructions eliciting a specific behavior pattern only once in every seven generations seems unlikely. A simpler explanation is that the animals are genetically programmed in some way such that reproduction commences only when certain lunar and solar photoperiodic conditions coincide. The same underlying genetic and physiological mechanisms would thus govern both three-year and twenty-eight-year spawning rhythm corrections.

Dan and Kubota (1960) provide an example of intercalation by a marine animal whereby certain lunar and solar photoperiodic conditions must both occur within an unusually narrow time limit for spawning to commence. *Comanthus japonica,* a crinoid or feather star, spawns once a year, sometimes on the third lunar quarter, sometimes on the first quarter, so that spawning intervals of twelve and twelve and one-half synodic months alternate regularly. Spawning is thus regulated so that it occurs within a period of about sixteen days on the Gregorian calendar. It appears that in the occasional years when the appropriate lunar quarter does not fall within this period the animals do not spawn!

1884; Harry, 1953; Ottino and Plessis, 1972; Titcomb, 1972). Brief references to the phenomenon have also occasionally been made by biologists (e.g., Billings and Munro, 1974; Harmelin-Vivien, 1977; Villadolid, 1940). As yet, however, we know very little about this subject. But it is evident from these observations (and those on spawning migrations) that adult reef and lagoon fishes are not all as restricted in their movements as has often been assumed.

Daily Rhythms

The timing of certain daily rhythms in fish behavior and movement can be predicted reliably from the phase of the moon through its influence on the timing of the tides. For example, many fish move from deep water up the reef slope and onto the reef flat to feed as the tide rises, then move back down again as the tide recedes. Others move from mangrove channels or depressions onto the reef flat on a rising tide. Their tidal migration routes are frequently narrow and well defined. Sand channels cutting through the reef edge, called *rams* in Palau, are favorite pathways for these movements, appearing like underwater highways at rush hour as fish stream through when the water reaches a certain height on a rising or dropping tide. Other species choose more subtle depressions in the reef as their pathways. Spears, hooks, traps, and cast nets can all be used to advantage here at these times.

During the first few years of commercial fishing (see chapter 5) the right to set traps in these tidal migration channels was hotly contested in Ngeremlengui. By 1974 catches had dwindled to the point where only the most productive *rams* were still in use. Even these were used only sporadically and provided only subsistence level catches.

On rising tides reef fish are often seeking food and thus less likely to seek shelter and enter traps than they are on falling tides, according to fishermen. (Fish on spawning runs reportedly seldom enter traps.) Some species congregate around the outer entrances of channels through the reefs to feed on what the outgoing currents bring. Line fishermen therefore work these areas mainly on the falling tide.[8]

8. A valuable description of the influence of tides on artisanal fishing is provided by Cordell (1974) for the estuarine canoe fishermen of northeastern Brazil. His account differs from this one in several ways: (1) because the fishery he described is not motorized, the fishermen are more severely restricted in their comings and goings by the tides. (2) The fishery occurs in a topographically very complex estuarine area; the consequent differences in tidal lags from place to place result in a more complex system for determining optimum fishing times on different fishing grounds. (3) Because net fishing predominates in this Brazilian setting, considerations relevant to spearfishing and line fishing are unimportant. (4) Cordell considers mainly the direct physical effects of moon and tide on fishing; he does not describe the diurnal and lunar rhythms in fish movement and behavior that may influence fishing schedules, although he acknowledges their probable importance.

Trolling outside the reef is better on a rising tide, particularly during the later stages, according to Palauan and Gilbertese fishermen and Korean commercial tuna fishermen with whom I talked. Japanese biologists studying commercial tuna catch statistics in Palauan waters reached a similar conclusion ("Report of Survey of Fishing Grounds and Channels in Palau Waters, 1925–26," 1937). None of these observers could provide a persuasive suggestion concerning the advantage to a fish of such behavior, nor can I.

There are two high tides and two low tides per day in Palau and their timing varies predictably with the moon's phases. Traditional Palauan fishermen commit this relationship to memory. On the day of the new moon, low tides always occur around noon and again about midnight, whereas high tides occur at about 6:00 A.M. and 6:00 P.M. Each day tidal extremes occur on the average of forty-eight minutes later. Thus, by the time of the next full moon, fifteen days later, the tides will fall twelve hours later. Because successive tidal extremes occur about six hours apart, this results once again in high tides at around dawn and dusk and low tides at about midnight and noon.

If a prolonged overcast prevents a fisherman from looking at the moon, he can determine the day of the lunar month by observing the times of tidal changes. Ngiraklang could time these changes precisely using a bit of nature lore. Near slack tide, current movement in the tidal creek at Ngeremlengui is too slight to be detected visually. So he looked into the water at a common little estuarine cardinalfish called *sebus 'l toach*. This fish, in the jargon of the marine biologist, is positively rheotactic; it faces into the current. Like many fish it can detect weaker currents than a person can perceive. Thus the upstream or downstream orientation of this fish told Ngiraklang whether or not the tide had turned. A century ago this trick would work only on a day when the sun was visible so that the time of day could be judged accurately by its height. Now Ngiraklang used a wristwatch.

The heights of high and low tides are also correlated with the moon's phases. The greatest tidal fluctuations (spring tides) occur on the days of full and new moons and the two or three days following each. The smallest tidal fluctuations (neap tides) occur about halfway between the spring tides, around the second and fourth quarters. The *kesokes* net is best used during the last two or three hours of a falling spring tide[9]; the water gets so low that most of the reef flat at

---

9. It is not possible to make a reliable list of the best days of the lunar months for kesokes fishing; there is too much month-to-month variation in tidal heights. In addition some nets are deeper than others and can thus be used during periods of less extreme low tides.

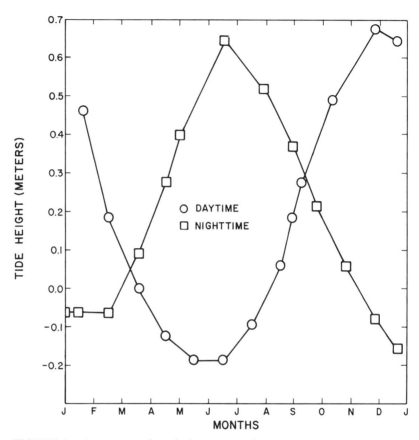

FIGURE 1. Lowest predicted daytime and nighttime spring low tides, Palau, 1976.

Ngeremlengui is drained, and fish trying to retreat to deeper water as the tide recedes find themselves trapped behind the net. During neap tides the water does not drop far enough to bare the reef flat and fish can evade the net. (Fishermen say that certain fish, such as *Lethrinus ramak (udech)* and *Lethrinus microdon (mechur)*, seem to be able to anticipate the degree to which the tide will drop; during neap tides they often sleep in shallow areas that are uncovered during spring low tides.)

The timing of *kesokes* fishing is also influenced by the fact that one of the two daily low tides is lower than the other. For six months (from roughly October through March) the lower of these two tides occurs at night. From roughly April through September it occurs during the day (figure 1). The difference in height between two low tides on a particular day may be as much as 0.85 meters. The *kesokes* net can only be used during the lower of the two low spring tides.

Thus for six months *kesokes* fishing is carried out mainly at night, for the other six months mainly during the day.

Catches made with *kesokes* nets are generally greater at night and are composed largely of different species than those caught during the day. At Ngeremlengui the rabbitfish, *meas*, and the emperor, *itotch*, are day feeders and are caught mostly during the day. Night feeders on the reef flat, such as the emperor, *eluikl*, and porgy, *besechaml*, form the bulk of nighttime *kesokes* catches, tending to stay in deeper water on the reef slope during the day.

Some trolling is done at night. One species, the barracuda, *Sphyraena genie (meai)*, is caught mainly at night. (Fishermen state that a piece of fish works much better as a nighttime trolling lure for *meai* than conventional feather lures, giving rise to the belief that *meai* must have good sense of smell.)[10]

Reef gleaning, the gathering by hand of small fish and invertebrates, is done mostly during the day. It is restricted largely to the season of very low daytime spring tides when most of the reef flat is uncovered. Fortuitously such tides occur during most of the *ngebard* season when offshore fishing is poor at Ngeremlengui and along the rest of western Babeldaob. So the harvest from reef gleaning by women and young boys provides a valuable supplement when the men make poor catches. Also more productive during these months was the traditional use of fish poisons. They were seldom used during *ongos* months because the daytime low tides were not very low (table 2); the water on the reef flats tended to be so deep, even at low tide, that poison was diluted rapidly and its effectiveness reduced.

Convenient days for dropline fishing are those of *mengeai*, the period of neap tides, the several days bracketing each of the second and fourth quarters. Because tidal fluctuations are smallest during those periods, tidal currents are at their weakest. The fisherman therefore has maximum control over his line. Less weight is required on the line to get the bait down, and it is easier to place the bait in a desired spot. The daily period when line fishing is easiest is for an hour or so during slack tide, when tidal currents are at their weakest. Fish tend to bite better, however, in an incoming tide; and, as is the case with many fishes throughout the world, early in the morning and again in the evening. With these three diurnal factors influencing

10. Until recently biologists thought that fast-moving pelagic ocean predators such as tuna and barracuda were visual feeders for whom smell played little, if any, role in detecting prey. Now, however, it is known that tuna have well-developed nasal capsules and can detect very low concentrations of dissolved olfactory substances (Atema, 1980). Although no research has been done on the chemical senses of barracuda, they too have very well-developed nasal capsules with many folds of olfactory epithelium (Bardach, pers. comm.). In short, the fishermen are probably right, once again apparently anticipating scientific observations by many years.

line fishing success, it is difficult to generalize concerning what times of day yield the best catches.

In reef and lagoon waters the clearest conditions usually occur during the first two or three hours of the incoming tide. At this time the underwater spearfisherman can scan a greater volume of water in search of fish. Fortuitously, this is also the time when many fish move from drop-off regions to shallower and more accessible reef areas. Neap tides also offer good conditions for spearfishing. As a consequence of reduced tidal currents, visibility is usually fairly good. In addition, less effort is needed to fight the currents. Aiming a nine-foot fishing gun accurately requires a steady arm and body, a requirement more easily met when turbulence is at a minimum.

During the day divers search as far as they can see through the water for fish. At night, in contrast, most hunting is done at close range in and around holes and crevices where fish rest. Water clarity and turbulence are thus less important at night. As noted earlier, spearfishing is much more productive at night because many species are inactive and easily approached. In addition, certain species that are active at night appear immobilized by the beam of a diver's underwater light.

The brightness of the moon is of great importance in nighttime underwater spearfishing. Reef fishes, say Palauans, feed more and rest less on bright nights (see also Helfman, 1978). Being more alert they evade spearfishermen more easily.[11] Bright nights occur from about the seventh to the twenty-second of the lunar month, that is, from half-moon waxing to half-moon waning. On the seventh the moon rises at about noon and sets about midnight, leaving six early morning hours of darkness and superior spearfishing. By the fifteenth (full moon) the moon rises at dusk and the entire night is bright. By the twenty-second the moon rises at about midnight, leaving six early evening hours of superior spearfishing.

For several nights beginning shortly after the new moon bumphead parrotfish, *Bolbometopon muricatus (kemedukl)*, rest in water as little as two feet deep at low tide. (This was volunteered by fishermen from Yap and Ponape in addition to Palau.) At these times nighttime spearfishing is particularly productive for this large and highly valued species. (As mentioned earlier, however, heavy fishing pressure seems in recent years to have brought about a shift of its main resting places to deeper water.)

11. The natives of Rangiroa similarly report that reef fishes are more active on moonlit nights (Ottino and Plessis, 1972). Hobson (1965, 1972a) has described how various tropical marine fish in Mexico and Hawaii also are more active on bright nights. Similarly Aristotle mentioned that in the Mediterranean, shoal fish travel at night only when the moon is shining. However, decapod crustaceans, including reef crayfish (i.e., "spiny lobsters") of the genus *Panulirus* are more retiring on moonlit nights (e.g., Allen, 1966; Sutcliffe, 1956)—perhaps in response to their greater visibility to nocturnal predators.

Nighttime trolling is more productive on moonlit nights (the tenth to twentieth lunar nights) when the lure is more readily visible. But fish do respond to trolling lures even on moonless nights. On such nights Palauans use larger lures to help compensate for the limited visibility provided by starlight. Although the human observer looking down into the water from above on a moonless night can usually see nothing, there is sufficient starlight so that a diver looking up can easily see fish silhouetted above him in shallow water (e.g., Hobson, 1966). Presumably, nocturnally active predatory fish similarly use starlight to help locate their prey.

In channels and along the outer reef slope large fish of many predatory species that stay in deep water during the day move up into shallow water and within easier reach of dropline fishermen at night. Fishermen in Palau, as in many parts of the world, take advantage of the fact that many species are attracted to artificial light on dark nights. Today dropline fishermen take pressure lamps out with them on such nights. Before the advent of lamps, palm leaf torches were used. This technique does not attract fish effectively when the moon is bright because its light tends to drown out that of a lamp or torch.

Bioluminescence is common in the lagoon at Ngeremlengui and provides a second reason for using a light on dark nights. Numerous small planktonic organisms will bump into a fishing line as it hangs in a current. This disturbance causes some of them to light up momentarily, making the line appear to glow. Apparently mistaking the glowing line for food, barracuda will strike at it, often severing it with their sharp teeth. Because this bioluminescence is relatively weak, the light from a pressure lamp tends to drown it, thereby reducing the threat from barracuda on dark nights. On bright moonlit nights the moon drowns out the bioluminescence. This phenomenon is much less often a problem in waters outside the reef where bioluminescence is less common.

In addition to daily tidal migrations, many reef fishes make well-defined migrations unrelated to moon phase at dawn and dusk. Most herbivorous reef fish and some carnivores feed during the day and rest at night (e.g., Helfman, 1978). Their feeding grounds are often some distance from their rest areas. Many carnivorous species stay in quiescent schools near the reef or seek shelter among corals during the day but leave the reef to forage for food in the water column, in seagrass beds, or over sand at night. Thus nocturnal carnivores return to the reef at dawn at about the same time that diurnal herbivores and carnivores move from their resting places to their feeding grounds. At dusk the reverse movements occur. Once again the migration routes of many of these fish are well known and are taken advantage of by Palauan fishermen. Some of these migra-

TABLE 3.   Preferred Lunar Phases, Tidal Stages, and Seasons for Fishing at Ngeremlengui*

	Best Lunar Stage	Best Tidal Stage	Best Season
KESOKES			
Daytime	Around new and full moons	Last hours of falling spring tides	April–September
Nighttime	Around new and full moons	Last hours of falling spring tides	October–March
UNDERWATER SPEARFISHING			
Daytime	Around the second and fourth quarters	Neap tides. Water calmest at slack tide, clearest during early stages of incoming tide	October–March
Nighttime	More productive than daytime. 23rd to 30th, 1st to 6th (dark nights)	Easier when tide is low	October–March
DROPLINE FISHING			
Daytime	Around 2nd and 4th quarters	Around slack and early part of incoming tide	October–March
Nighttime	23rd to 30th, 1st to 6th (dark nights)	Hard to generalize (see text)	October–March
TROLLING			
Daytime	Hard to generalize (see text)	Rising tide	October–March
Nighttime	Moonlit portions of 7th to 22nd		October–March
REEF GLEANING			
Daytime	Around new and full moons	Spring low tides	April–September
PORTABLE TRAPS	no preferred time		
CAST NETTING	The best lunar phase, tidal stage, and season vary with the species being sought.		

*Superimposed on these determinants of good fishing are the spawning rhythms and diurnal migrations of the fish.

tion pathways coincide with those of tidal migrations described above. Other pathways are restricted to the reef flat and have no vertical component.

A number of biologists, most notably Edmund Hobson, have described similar twilight movements in other coral reef communities: "This activity involves movements that occur consistently in certain locations . . . specific routes are followed, in which certain species stream continuously past a given point in long drawn-out processions over a period of several minutes" (Hobson, 1972*a*, p. 720). He says elsewhere: "Even species that are solitary when their activity is confined to a limited area will join with others of their kind when migrating" (Hobson, 1973, p. 367).

Turtle Rhythms

Hawksbill and green turtles are still common in Palauan waters, even though they have been heavily fished. Nevertheless, say fishermen, their numbers and average size have decreased noticeably. Hawksbills nest in Palau, particularly in the Seventy Islands area, where they crawl ashore at night to lay their eggs in the sand. (A few green turtles nest on small islands north of Babeldaob and on Peleliu, but the major nesting sites in the Palau district are Helen Reef and Merir, small uninhabited islands south of the Palau archipelago.) Until the Seventy Islands were declared a marine reserve a few years ago, fishermen came to harvest hawksbill eggs for food and the adults for their shells, which were made into "women's money" and jewelry. In recent years Palauans have learned that the strong-smelling meat of the hawksbill becomes palatable when it is boiled several times, changing the water each time. Consequently the flesh is not wasted as it once was.

The nesting season of hawksbills in Palau stretches from June to January, with a peak, according to fishermen, in July and August. Like turtle fishermen throughout the tropics, Palauans are familiar with the fact that female turtles lay their eggs several times per season and individuals will usually return in the evening on a high tide to the same stretch of beach for each laying. It has been suggested that turtle egg harvesters probably originally discovered these facts by noting that individual turtles with distinguishing marks, such as scars, returned to the same beach to lay their eggs at regular intervals (Carr, 1972).

This may well be true, but Palauan legend records a more entertaining (and not entirely implausible) version of how the discovery was first made in Palau. Long ago during the time of the gods a young man and a girl fell in love. As the distance between their home islands was great they decided to rendezvous on Ngemelis Island which lay between them. They first met on the night of the new moon. When they awoke in the morning the girl could not find her grass skirt, and was forced to make a makeshift skirt from palm fronds. Before parting, the lovers agreed to meet again fifteen days later on the following full moon.

They arrived on the appointed evening, and as they lay by the beach they saw by the light of the full moon a turtle crawling toward them. Fragments of a grass skirt were entangled around one flipper. It was then that they deduced the fifteen-day egg-laying cycle of the hawksbill and the fact that individuals nest repeatedly on the same beach. (This could have been, inadvertently, the world's first re-membered marine tag-and-recapture experiment!)

Palauans state that more turtles lay their eggs around new and full moons (that is, on spring tides) than at other times, although some egg-laying occurs throughout the lunar month. Turtle biologist D. C. Drummond similarly observes (personal communication) that more green turtles beach at Heron Island, Great Barrier Reef, on spring tides than at other times. He points out that as these turtles have an egg-laying cycle of about two weeks, and as they can delay laying in the presence of unsuitable conditions (such as insufficient water over the reef flat at high tide to facilitate their reaching the beach), a tendency for egg laying to synchronize with spring tides is not unreasonable. Hocart (1929) reported that turtles in Fiji lay their eggs largely around the full moon. As nesting there presumably also occurs at approximately two-week intervals, laying peaks therefore should also occur around the time of the new moon.

Palauans have taken the ability to predict when a turtle will return to its nesting beach two steps further. They have learned that by examining the eggs they can deduce how long ago they were laid. And if the eggs are less than fifteen days old they can estimate how many days will elapse before the parent returns to lay the next batch.

An egg, when laid, is rubbery and flesh-colored. But the shell begins immediately to calcify and harden. Calcification begins as a white disc that gradually enlarges and spreads over the entire shell. Up until the sixth day after the eggs have been laid, the experienced turtle hunter can estimate their age by the size of the calcified region. After the sixth day an egg must be peeled open and the size and state of the developing embryo used to determine the age of a nest. By the fifteenth day, for example, the umbilical cord is clearly visible. Using a piece of twine the fisherman ties a number of knots equal to the calculated number of nights that will elapse before the turtle will return to lay its next batch of eggs. By removing one knot each day he knows when it is time to intercept the turtle on its return to the beach.

This technique is not perfect because the fifteen-day egg-laying cycle is only approximate; the female may return on the fourteenth, or more rarely the sixteenth day. In addition, according to fishermen, the embryos do not develop at exactly the same rate, growing more slowly in shaded or overly moist nests or in rainy weather. (Too much fresh water collecting in the nest is liable to cause the eggs to rot, they say. Roots growing around the eggs will hinder the escape of hatchlings.[12]) Solomon Islanders apparently used similar criteria to determine when turtles would return to lay their next batch of eggs (Hocart, 1929).

A second observation allows the turtler to distinguish between an individual turtle's first clutch of eggs for the year, its last clutch, and intermediate layings. The eggs at the bottom of the first clutch are small, elongate, have little yolk, and seldom hatch. There are few misshapen eggs in the intermediate clutches. In the last clutch the eggs on *top* of the clutch are small and misshapen. It is as if the reproductive machinery of the turtle is a little rusty early in the season and falters once again just before it shuts down at the end of the season, producing inferior eggs in both instances.[13] Apparently the Polynesians of the Tuamotus are also aware of this phenomenon; Emory (1975, p. 217) states, "the last eggs to be laid were smaller than the others and were called *teke titi*. When such eggs were observed it was a sign that the turtle would not come ashore again that season."

Among a few Palauans a curious belief exists that the number of eggs in a nest will reveal how soon the female will return to nest again. Eighty-four eggs would mean four days, eighty-seven eggs would mean seven days and so forth. This notion can also be found among the natives of Truk (LeBar, 1952), the Tokelaus (Beaglehole and Beaglehole, 1938), and Tuamotus (Danielsson, 1956).

Turtles feed mostly during the night, early morning, and late afternoon. Often around midday they move into the lagoon and sleep on the bottom for two or three hours. The hawksbill generally sleeps in a crevice or cave in the reef; the green turtle more often chooses a sandy bottom, under an overhanging coral or rock. Both species also sleep during part of the night, hawksbills generally sleeping longer than green turtles. Both have customary sleeping places with which Palauans are familiar. The animals are easy to catch here because they are almost oblivious to disturbance. Palauans say a person who is hard to wake up *bad el wel*—"sleeps like a turtle."

12. For a more detailed account of the various hazards faced by turtle eggs according to Palauan fishermen, see Helfman (1968).

13. The first and last eggs laid by certain geckoes during their reproductive lives are similarly small and misshapen (Falanruw, pers. comm.).

C H A P T E R

SEABIRDS AS FISHFINDERS 4

A striking scene is played out many times each day in Palauan waters as schools of forage fish try frantically to elude large tuna or other predators. Periodically wave upon wave of small silvery bodies erupt from the water then shower back, making a sound like hard rain. Here and there the water, seems to boil where pursuers slash through the school. In moments the air is filled with fluttering wings and strident cries as seabirds converge to join in the carnage.

On clear, calm days the commotion created by the fish may be spotted by fishermen from a distance of a few hundred yards. This range is greatly reduced if the water is rough. But the accompanying seabird flocks can be seen for thousands of yards even in rough weather. Furthermore the birds will spot and follow schools of fish well below the surface and quite invisible, even at close range, to fishermen. So for many centuries the fishermen of Oceania have relied on birds to help them locate fish.

Today American and Japanese skipjack fishermen in large vessels fitted out with modern electronic equipment have scarcely improved on this method of fish finding. Binoculars have extended the maximum range over which seabird flocks can be seen up to eleven or twelve miles (Orbach, 1977). (From the bridge of a tuna boat the practiced, unaided eye is good to about three miles in clear weather.) But otherwise the method remains unchanged.

The behavior of the birds as they feed or search for food is tuned finely to the behavior of the big fish. Different predatory fish differ in the speed and pattern of their search and feeding movements, and these differences are mirrored in the movements of the birds above them. Different species of fish also select different size ranges of prey, and so do the different birds. A bird tends to be found over those predatory fish whose food-size preferences tend to match his own.

By watching the birds, consequently, the skilled fisherman can not only tell where fish are but can also obtain clues concerning their identity, whether or not they are feeding (and thus liable to take his lure), and, to a limited degree, what they are feeding on (indicating the kind of lure he should use). The unskilled fisherman pursues any flock he sees. The skilled fisherman chooses his flocks. Prospects are best when large numbers of birds are densely concentrated very near the surface and dive fast and often. The size and spread of a feeding flock is an indication of the number of fish feeding beneath it.

Although Ngiraklang was mainly a reef and lagoon fisherman he sometimes trolled outside the reef. Through watching and talking with other fishermen he developed considerable skill in reading seabird behavior. About one-half of what he told me about Palau's seabirds can be found scattered in the scientific literature on seabird behavior.[1] In no instance did his observations contradict those already published by ornithologists. This, plus my own observations, made during a commercial tuna fishing trip I took after talking with Ngiraklang, gives me added confidence that the additional information he volunteered is accurate.

The most abundant seabird in Palau is the black noddy tern, *Anous minutus (bedaoch)*. Flocks are commonly seen quartering back and forth a few feet above the surface in search of feeding schools of fish. If Ngiraklang sees them feed for a while, while moving fairly rapidly across the water, then fly quickly off a hundred yards and begin feeding again, he suspects that skipjack tuna, *Katsuwonus pelamis (katsuo)*, are feeding beneath them. These fish move rapidly while diving and change direction frequently, surfacing

1. A good source of information on the comparative feeding ecology of tropical seabirds, including two of the three species discussed here, is Ashmole and Ashmole (1967). Other sources of information on the feeding behavior of tropical seabirds and the fish they compete with are Murphy and Ikehara (1955), Royce and Otsu (1955). Papers by Nordhoff (1930) and Anderson (1963) on Tahitian offshore fishing lore are also useful. However the fish that these last two authors referred to as bonito are in fact skipjack. What Nordhoff refers to as albacore are actually large tuna of several species, albacore being one of them. As Anderson points out, some of Nordhoff's bird identifications were also incorrect.

to feed periodically. The deeper the fish dive the higher up the birds fly above them.

In the presence of feeding yellowfin tuna (*tkuu* or *manguro*) *Thunnus albacares*, noddies behave much as they do over skipjacks. But yellowfin move somewhat slower than skipjack and change direction less frequently. This is reflected in the movements of the birds above them. Yellowfin also break the surface more forcefully and often than skipjack, frequently leaping completely clear of the water. Schools of yellowfin and the flocks of birds that follow them are often separated into several adjacent subgroups.

A flock of noddies moving very slowly and steadily over water churned up more heavily than it would be by feeding skipjack signals the presence of kawa kawa, *Euthynnis affinis (soda)*. Sometimes rainbow runners, *Elagatis bipinnulatus (desui)*, travel in mixed schools with kawa kawa but seldom break the surface as they feed.

Wahoo, *Acanthocybium solandri (keskas* or *mersad)*, seldom travel in compact schools. A few birds, diving sporadically to the surface, signal their presence, or that of the great barracuda *Sphyraena barracuda (chai)* or Spanish mackerel *Scomberomorus commersoni (ngelngal)*—both are usually solitary feeders. Ngiraklang said that these fish feed on relatively large forage fish and that the noddies wait until they see the fish chasing prey small enough for them to eat before they dive.

If birds are feeding a mile or more from the reef, then barracuda, wahoo, and Spanish mackerel can be ruled out because they rarely venture that far offshore. If birds fly high and slowly without diving they are probably following a school of tuna travelling in deep water, waiting for it to surface and feed. One or more birds flying slowly up and down over the same spot suggests that a floating log is beneath. Such logs generally signal the presence of large numbers of fish below (see chapter 7). Flocks of black noddies feeding within Palau lagoon usually signal the presence of jacks or kawa kawa.

When a flock of feeding noddies breaks up this indicates that the fish have dived. The fisherman then looks higher in the sky in search of white terns, *Gygis alba (seosech)*. When looking for food these birds, which usually search singly or in twos or threes, stay higher above the water than noddies or shearwaters. Because of their superior vantage point they can spot feeding fish from a greater distance and in deeper water than can the latter and are thus often the first birds to reach a school.

White terns flying in relatively low circles, making sudden fast dives to feed, indicate the presence of yellowfin tuna. White terns flying high, then low and feeding wildly and erratically, often in the absence of other birds, signals to Ngiraklang (as well as fishermen from Tahiti to Hawaii to Japan) the presence of the dolphin-fish *Coryphaena hippurus (chersuuch)*.

Audubon's shearwater *Puffinus lherminieri (ochaieu)* usually flies only about a foot above the surface, and ranges much further out to sea than noddies or white terns. This bird prefers comparatively small fishes that are fed on largely by skipjack or kawa kawa. Feeding *ochaieu* are thus usually a sign of the presence of either these fishes, or of small yellowfin tuna.

The shearwater can sit on the water as it feeds and, unlike the terns, is able to stick its head well beneath the surface in pursuit of fish, although it rarely submerges its body. Noddies and white terns feeding on the wing will grab several fish at a time, often subsequently spilling some of them into the water. The "lazy" shearwaters sitting below pursue these dazed or injured fish.

I asked Ngiraklang if he could explain how it was these three species of birds could coexist while feeding on what appears to be much the same food: "Why doesn't the most aggressive species outcompete the others and gradually replace them?" This kind of question is often raised by biologists interested in competition between different species, but is often not easily answered. Ngiraklang answered without hesitation.

The white tern begins to hunt earlier in the morning than the other two birds, he pointed out. Its greater cruising altitude also enables it to find and reach feeding schools faster than the other birds. (The other birds, in fact, use *it* to help them find feeding schools.)

The noddies are the largest and most aggressive of the three species. Once they spot a school of fish on which white terns are feeding they tend to crowd in beneath them and usurp the best feeding positions. They also tend to feed closer to shore than the white terns although there is considerable overlap.

Audubon's shearwater can sit on the water and partially submerge, unlike the two other birds. It can therefore catch fish inaccessible to its competitors. And, although there is an overlap in size preferences, the shearwater tends to focus on smaller fish than the other two species. Furthermore it ranges much further out to sea.[2] Thus each bird has a "corner" on different portions of the shared resource.

2. Native navigators in the Pacific islands use noddy and white terns as indicators of the presence of land within about twenty miles. Shearwaters, they know, range too far from land to be useful to them in this connection (Gladwin, 1970; Lewis, 1972).

CHAPTER

THE TRADITIONAL CONSERVATION ETHIC AND ITS DECLINE

5

Wisdom fails in the face of radically altered conditions.
—*source unknown*

It was not until this century that Europeans and Americans began to realize that the oceans' yield of seafood has practical limits. Previously our wide, productive continental shelves, low population densities, and great terrestrial food resources kept us from exceeding and thus perceiving these limits. Most Pacific islanders, in contrast, lack important terrestrial sources of animal protein and their vital marine food supply has been limited largely to the narrow strips of shallow coastal waters around their islands. Consequently they have long been aware of the value of husbanding their marine resources, and, centuries ago, devised and employed almost all the marine conservation measures we continent dwellers developed only recently (Johannes, 1978a).

In Palau, however, seafood seems always to have been abundant. There are no suggestions in Palauan folk history or legend that food getting was ever a problem. This is probably due in part to the fact that the Palau archipelago possesses an unusually high proportion of sheltered, productive reefs to land. In addition, chronic warfare seems to have helped keep the population below the reefs' carrying capacity. Nevertheless Palauans possessed a well-developed marine conservation ethic. *Bul*, a term roughly equivalent to "conservation laws," were manifest in a variety of traditional controls on reef and lagoon fishing. It is necessary to look beyond biological limits on seafood production to find the explanation for their existence.

One result of warfare was that until this century Palauan villages were situated inland in order to afford their inhabitants better protection from enemies. It was dangerous in times of war to be caught alone far from the village. Thus fishermen tended to fish within a limited distance of the adjacent coastline. In addition, bad weather discouraged fishermen from venturing far from shore even in times of peace. (Even today fishing boats stay close to home for long periods during the season of onshore winds.) Although seafood was abundant in Palau as a whole, nearshore stocks were therefore probably heavily exploited. Political and meteorological conditions rather than the total carrying capacity of Palau's reefs thus appear to have provided the main incentives to develop marine conservation measures.

The most important form of marine conservation used in Palau, and in many other Pacific islands, was reef and lagoon tenure. The method is so simple that its virtues went almost unnoticed by Westerners. Yet it is probably the most valuable fisheries management measure ever devised. Quite simply, the right to fish in an area is controlled and no outsiders are allowed to fish without permission.

Where such tenure of marine fishing grounds exists it is in the best interests of those who control it not to overfish. The penalty for doing so—reduced yields in the future—accrues directly and entirely to the fishermen owners. Self-interest thus dictates conservation. In contrast, where such resources are public property, as is the general case in Western countries, it is in the best interest of the fisherman to catch all he can. Because he cannot control the fishery, the fish he refrains from catching will most likely be caught by someone else. Self-restraint is thus pointless. Self-interest dictates overfishing and leads to shrinking yields.[1]

For centuries each of Palau's village clusters has exercised the right to limit access to its fishing grounds. Within at least one

1. The elucidation of this problem in general ecological terms is often associated with Hardin and his well-known article "The Tragedy of the Commons" (Hardin, 1968). In it he states, "Ruin is the destination to which all men rush, each pursuing his own best interests in a society that allows freedom of the commons." Economists recognized this problem long before this (as Hardin acknowledged). A classic description of the economic superiority of limited entry to, over public ownership of, a

district (Ngaraard) further subdivisions were made in this century so that individual villages have control of adjacent fishing grounds. These fishing rights are still maintained, in most municipalities, to just beyond the outer reef dropoff. (Up until the turn of the century, when the custom of shark fishing miles offshore died out [see p. 14], fishing rights extended in theory to the outer limit of foraging seabirds resident on Palau, or about 75-150 miles [Nakayama and Ramp, 1974]. The need to defend this vague outer boundary arose very rarely if ever.)

Today traditional fishing rights are still controlled by the chiefs for the benefit of the villagers they represent. A municipality concerned about poachers will sometimes broadcast a warning to them over the local radio station. Continued poaching can lead to the chief of a poacher's village being fined by one of Palau's traditional high chiefs. The fined chief thereby loses face. And the clan of the fisherman who caused this embarrassment is made to pay for it—usually in the form of a substantial amount of cash. Formerly hostilities between neighboring districts often precluded the observations of these niceties, and the offenders, if caught, forfeited their lives.

The system was not inflexible however. The use by others of marine resources surplus to the needs of the owners could be arranged. Fishermen were sometimes allowed to fish in their neighbor's waters providing they asked permission and agreed to pay a portion of the catch. (Today some districts forbid outsiders to fish commercially in their waters but allow visitors to fish there for their own needs.) Sharing of fishing resources sometimes went beyond temporary fishing permits to the outright gift of fishing grounds to less well-endowed villages. For example, in about 1930 the municipality of Ngeremlengui ceded fishing rights to two reef and lagoon areas surplus to their needs to the neighboring district of Ngatpang.[2]

---

renewable resource was published by Gordon (1954). He focused on this problem specifically as it relates to fisheries management. Today most Western fisheries biologists and economists consider limited entry, or equivalent measures for managing a common-type resource, to be the sine qua non of good fisheries management (e.g., Christy, 1969; Crutchfield, 1973; Nielsen, 1976; Van Meir, 1973), and are slowly managing to persuade their governments that the painful changeover from open to limited access must be made in order to revitalize their fisheries.

It could be argued that reef and lagoon tenure might have been devised originally more as a means for decreasing conflict on and providing equitable access to the fishing grounds than as a conservation measure. This may be true although we will never know for sure. But regardless of its original purpose, reef and lagoon tenure functions as an important conservation measure, and Micronesian fishermen today are well aware of this. (I found this to be the case even in Ponape where colonial interference led to its destruction three generations ago.)

2. Until early in this century Ollei, Ngeremlengui, and Koror shared the fishing rights along the entire sixty-mile long barrier reef of western Palau. During the Japanese occupation a head tax of five yen was levied on Palauans. One of the few

Fishing grounds may be shared by two districts. Traditionally, for example, Kossol and Ngerael reefs have been jointly fished by the people of Kayangel and Ngerchelong. Fishermen from Ngeremlengui and Ngardmau may exploit one another's trochus resources commencing three days after the opening of trochus season (Kaneshiro, 1958).

Another useful form of conservation was simply to avoid waste. Apprentice fishermen were taught that it was bad to catch more fish than could be consumed. Old men I asked about traditional conservation customs often volunteered, "In the old days we would only take the fish we needed to eat from the *ruul* and let the rest go." But because it was easy to underestimate the catches of others in the village, fishermen sometimes inadvertently kept more fish than could be used immediately. Such fish were not wasted, however. Some were smoked and thereby preserved for a day or two (Keate, 1788). Fish stews were made and reheated once or twice daily, sometimes for weeks, to prevent spoilage (Kramer, 1929).

To ensure a supply of easily accessible food during bad weather, laws were made that restricted the use of certain species to such times. On the island of Peleliu there is a large brackish pond in which milkfish live. Traditionally the harvesting of these fish was forbidden except during stormy weather. Even today in Ngeremlengui sea cucumbers and giant clams are seldom eaten during the good weather months of *ongos* in a conscious effort to conserve their populations for use during the rough *ngebard* months when fishing is poor.

Derris root (*dub*) and other poisonous plants were used to poison reef fish (e.g., Abe, 1938). It was recognized that too much *dub* applied in the vicinity of corals would kill them—and that when the corals were dead the number of fish in the vicinity would often diminish.[3] Consequently, according to Ngiraklang, young fishermen were taught not to tuck their bundles of *dub* under coral heads.[4]

---

sources of cash available to the natives was the sale of trochus shell, used by the Japanese to manufacture mother-of-pearl buttons and jewelry. Because trochus were obtained only on the outer barrier reef, Ollei, Ngeremlengui, and Koror shared a monopoly on the west coast supply. Other western villagers without this source of revenue were hard put to pay their taxes. Consequently the owners of barrier reef fishing rights were told by the Japanese to allocate portions of them to other villages. As a result the villages of Aimelik, Peleliu, Ngatpang, and Ngardmau came into possession of fishing rights along the western barrier reef, which they retain today.

3. Elsewhere I have observed a marked decrease in number of reef fishes in areas where corals have been killed (Johannes, 1975).

4. Hundreds of bundles of *dub* were used occasionally to collect fish for feasts, however (Kramer, 1929). This poison kills all sizes of fish and therefore undoubtedly took a large toll of juveniles.

In the village of Ollei it was customary to allow certain species of fish, particularly the rabbitfish, *Siganus canaliculatus* (*meas*), to mass on the spawning grounds for at least one day before pursuing them, thereby allowing a portion of them to spawn. Early in this century a chief in the village of Ngiwal similarly forbade catching some species of fish while they were spawning. (The vulnerability of spawning fish to Palauan fishermen is described in chapter 3.) These customs, like many others, were discontinued as a result of the disruptions of World War II.

An unusual conservation measure was once practiced in northeast Babeldaob. A type of herring, *Herklotsichthys punctata*, or *mekebud*, occurred in dense shoals at certain times of the year (Kramer, 1929). Jacks, particularly *Caranx melampygus* (*oruidl*), patrolled these shoals, often driving them into very shallow water. Frequently the *mekebud* would be driven right up onto shore in their efforts to escape[5] and here the people could readily gather them. Sometimes the *oruidl* also beached themselves as they chased their prey. Here, one might guess, was a double windfall as *oruidl* were also a favored food fish. It was the law, however, that stranded *oruidl* must be returned to the water so that in the future they might drive more *mekebud* ashore. (A similar traditional law exists in the Marshall Islands and is still in effect on Likiep Atoll according to Marshallese fishermen.)

The god of the small island of Ngerur, north of Babeldaob, owned the island's turtles. Consequently no turtle could be caught while on the island and no turtle eggs could be dug. In some other areas Palauans were not supposed to kill a nesting turtle until after it had laid several batches of eggs or until it had reached the water after nesting. When turtle eggs were harvested it was the law in certain parts of Palau that some had to be left to hatch.

The Impact of the Twentieth Century

Traditionally a Palauan's needs and wants were satisfied almost exclusively from within his internally regulated subsistence economy. He lived in a state of what Sahlins (1974) has described as "subsistence affluence." Because economic enterprise, according to Useem (1945, p. 570), "rated low in the scale of values, the accumulation of physical goods had no effect on a person's status, sense of security, or standard of living." And, as in much of Oceania, the Palauan worker had control of his own economic means. Sahlins

---

5. Palauan children sometimes catch *mekebud* by jumping from an overhanging mangrove branch into the midst of a school. The startled fish leap into the air and some of them land on the bank where the children scoop them up.

(1974) notes that this circumstance rules out elevating one's self through controlling the production of others, as is done in most industrial societies. "The political game," he states, "has to be played out on levels above production, with tokens such as food and other finished goods; then, usually the best move, as well as the most coveted right of property, is to give the stuff away."[6]

In accordance with this principle the traditional Palauan fisherman seldom sold his fish; he generally gave them away.[7] "To have is to share," is a traditional Palauan saying (Force, 1976). Ultimately some of the products of the recipients' labor were tendered in return.[8] Village social and economic life were in this way intimately entwined. One impact of foreign cultures on Palau has been to erode this system.

Japanese colonists imported a variety of trade goods and set up stores to provide Palauans with the incentive to work for them (e.g., Yanaihara, 1940). This planted the seeds of a new economic order, subsequently nurtured by the United States, whereby goods are seldom shared, but rather bought and sold.[9] The fisherman began to sell some of his fish to the Japanese in order to obtain the money to purchase these attractive new commodities. In order to catch enough fish to continue, in addition, to supply the needs of his village he had to fish harder. But among the newly imported goods were excellent nets, motorized boats, and a host of other items that made his fishing easier. The creation of export markets (e.g., Japan for trochus shell, Asia for sea cucumbers, and later, during the U.S. administration, Guam for reef fish) expanded his opportunities for earning money.

At first the reefs withstood his increased harvesting pressure because motorboats facilitated fishing in more distant areas seldom exploited in the days of dugout canoes and chronic warfare. But eventually his catches began to drop while imported inflation occasioned by Palau's increasing participation in external markets caused operating expenses to climb. Increasingly he found himself forced to compete with his fellow fishermen for money and thus for fish. He abandoned the leaf sweep, which employed a dozen or more

---

6. It is sometimes assumed that giving away the fruits of one's labor in a communal society is less egocentric than selling them in a capitalistic society. Such is not necessarily the case, as Sahlins clearly points out in this passage. Palauans do not generally strike the observer as being particularly altruistic.

7. There was some sale of certain highly valued marine animals, such as dugongs, by villages in whose waters they were found, to unendowed villages (Kubary, 1895).

8. One consequence of this traditional mode of distribution in Palau is that older Palauans often refuse to believe that there can be places in the world where people go without food simply for lack of money.

9. Unlike many Pacific islanders Palauans had a traditional form of money, but it was used mainly for ceremonial transactions (e.g., Force, 1976).

men who worked cooperatively and shared their catch, and he adopted the imported *kesokes* net, which he could operate by himself. He invested in progressively better and more expensive fishing gear and boats.

As fishing equipment became more sophisticated its price rose beyond his means, so he began to borrow money and found himself unable to give away as many fish in the village as he used to. He was developing "crippled hands" said the villagers. They grew less willing to share their goods and services with him and began to demand payment for them. To obtain the cash thus required he found himself compelled increasingly to sell fish within his own village. Where fish had been a form of social security they were now becoming a commodity. Social and economic systems were separating. The communal subsistence economy was now under stress not just in the main trading center of Koror but increasingly also at its base in the outlying villages.[10]

This disruption of adaptive fishing patterns and associated distribution patterns differed little in outline from that which has occurred throughout Oceania and elsewhere when money, new technology, export markets and the profit motive impinge upon communal fishing economies (e.g., Alexander, 1975; Cordell, 1973; Forman, 1970; Nietschmann, 1973; Porter and Porter, 1973). But examining some of the specific events that accompanied these changes in Palau is useful if we are to try to understand contemporary Palauan attitudes to their marine resources.

In a flush of enthusiasm over the wealth suddenly made available through the opening of an export market for reef fish to Guam in 1959, Palauan fishermen obtained U.S. government loans to purchase bigger boats. In an effort to keep up the payments on one large vessel the people of Ngeremlengui fished their reefs down to a catch level at which it no longer paid to keep the boat in operation and they had to forfeit it. Some of the old men who still remember when they gave away their then-surplus fishing grounds to other municipalities now rue the day. (They also resent the fact that the recipient communities no longer manifest any gratitude for the gifts. "In Ngatpang five chiefs have come and gone since we gave them some of our fishing grounds," said one old fisherman, "and now hardly anyone remembers.")

10. The socioeconomic changes described here were interrupted in World War II during which the reefs were very heavily fished to supply Japanese troops. By the end of the war the Japanese economy was totally destroyed and Palauans were forced temporarily back to ancestral living patterns (e.g., Useem, 1955). Reef fish populations temporarily rebounded according to fishermen. Within a few years, however, the process of monetization of the economy and the concomitant depletion of natural resources reasserted itself under the U.S. administration.

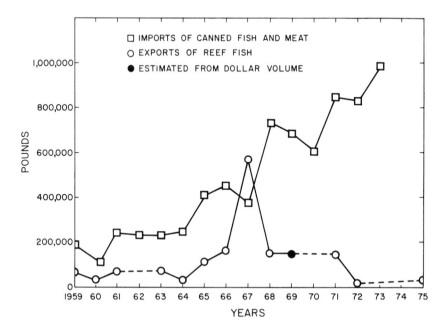

FIGURE 2.   Imports of canned meat and fish, exports of reef fish, Palau.

As fish populations dwindled in heavily fished districts such as Ngeremlengui and Koror, their fishermen began to eye the reefs belonging to others whose waters were not so heavily fished. Efforts (unsuccessful so far) were made to enact legislation invalidating traditional reef and lagoon tenure laws. The other municipalities became reluctant to grant fishing rights to what had become an important cash resource, even if they did not exploit it fully themselves. They saw what had happened to the fishing grounds of their ambitious neighbors and preferred to let their reefs remain underharvested rather than see them badly depleted.

Over the past decade many young Palauans have turned their backs on village life and flocked to Koror. Over 60 percent of all Palauans now live there. The waters nearby are overfished, and men with nine-to-five jobs do not have time to catch enough fish for their families. Most of the many unemployed have no boats and, in any event, look down on fishing as an occupation. Thus most residents of Koror now buy their fish.

By far the most popular imported animal protein is canned mackerel from Japan. Although Palauans do not think highly of its flavor, the economy of scale involved in its capture and processing has facilitated its sale in Palau at prices comparable to those of locally caught reef fish. Its ease of preparation and its indefinite shelf life also contribute to its popularity. Canned fish now also fill the gap when the weather is poor and fishing is unproductive in the outer villages. Each village has several small stores and, next to soft drinks, canned mackerel is the biggest seller.

When a family must purchase fish there is another incentive to choose canned fish; it stretches a lot further. "A pound of reef fish feeds one person," pointed out one informant, "a pound of canned fish feeds three or four people." The reason for this is that a few ounces of strongly flavored, oily canned mackerel provide a feeling of surfeit, when mixed with rice, that requires a much larger portion of reef fish to accomplish. (People in other parts of Micronesia also pointed this out to me.)

So dependent had Palauans become on imported animal protein that by 1973 a staggering eighty-two pounds of canned meat and fish[11] were imported for every man, woman, and child (figure 2). Nutritionally self-reliant fifty years ago, Palauans now imported almost one-third of the animal protein they consumed[12] (as well as large quantities of polished rice, refined flour, and sugar).[13]

Whereas 576,000 pounds of Palauan reef fish were exported at the peak of the export fishery in 1967, exports had dropped to less than one-tenth of this level by the early 1970s.[14] The weight of meat and fish imported therefore now greatly exceeded the weight of reef fish exported. Thus despite the fact that Palau's population was much lower than in times past,[15] and despite substantial improvements in mobility of boats and efficiency of gear, fishermen no longer harvested a quantity of fish equal to Palau's own needs.

A local tuna fishery was developed by a U.S. company but few Palauans chose to enter it. Most of the jobs thus fell to imported

11. The information is taken from the Palau District Administration statistics. In addition to this there was imported frozen meat and chicken (available in a number of Koror grocery stores over the past few years) for which no figures are available.

12. Palauans consume an average of about seventy-four to eighty-one grams of protein per day (Hankin and Dickinson, 1972), about 75 to 80 percent of which is animal protein (Hankin, pers. comm.). Assuming reef fish are about 27 percent protein (Murai et al., 1958) and that one-third of the landed weight of fish is not consumed (scales, bones, portions of the internal organs, plus portions fed to dogs, cats, and pigs), it can be calculated that the average Palauan consumes animal protein equivalent to about 270 pounds of fish in the round per year.

13. Nutritionally sound diets are often replaced with inferior ones as market economies replace subsistence economies (e.g., Hoyt, 1956). Cholesterol, blood sugar, and blood pressure levels increase progressively as one moves from Palau's remoter villages toward the district center (Hankin and Dickinson, 1972).

14. The information is taken from the Palau District Administration statistics.

15. At the time of first Western contact the population was estimated to be around 40,000-50,000 (Semper, 1873)—more than three times what it is today.

Okinawans and Koreans. It was said that Palauans made poor tuna fishermen because the long trips beyond the reef were culturally unpalatable. But an economic analysis of the fishery (Callaghan, 1976) showed that pay levels were not competitive. The U.S. government had raised the salaries of Palauan government employees, who constituted more than one-half of the employed work force, toward U.S. civil service levels, thereby accelerating imported inflation and pricing Palauans out of their own labor market.

Contemporary Attitudes

In the 1970s attitudes toward conservation differed significantly with age in Palau. Many of the old men I talked to were concerned about Palau's natural resources. They lamented the declining numbers of fish and pigeons, and the shrinking forests, all of which they attributed to overharvesting. A number of these men, including Ngiraklang, were chiefs. But their power to enforce Palau's traditional laws had been greatly eroded by successive colonial governments.

In 1955 at the behest of the U.S. administration a unicameral legislature was chartered. Elected members from the various municipalities had voting rights. But hereditary chiefs were relegated to a separate table and had no vote. Initially the aims of the elected members and the chiefs seemed reasonably congruent; Palauan leaders have traditionally been less conservative than those in many other Pacific island cultures (McKnight, 1972), and both chiefs and elected members pursued the adoption of those Western customs they perceived as promoting constructive change.

But soon it became apparent that the actions of the elected members, many of whom were businessmen, were often not in Palau's best interests. As early as 1949 Useem had noted, "Palauan businessmen take no active part in issues not directly related to their own practical interests." Many of the elected members were educated in U.S. institutions in the 1950s and early 1960s before the image of the American businessman had lost much of its glamour. The successful businessman rather than the wise chief became the ultimate object of admiration and emulation; making money for one's self became an overriding concern.[16] Palau's communal traditions stood in the way and were increasingly ignored except when they served conveniently as a form of political patronage. The political game no longer consisted of "giving the stuff away" but of

16. A survey carried out in 1955 in the outlying village of Ngerchelong showed that children would much rather become storekeepers than chiefs (Force and Force, 1972)—a striking indication of the influence of the emerging market economy on Palauan values.

investing it—in projects that often bore no relation to the welfare of one's fellow Palauans. The chiefs became increasingly troubled by the actions of the elected members but could do little about it. The U.S. administration dealt primarily with elected members and tended to ignore the chiefs, who "were accorded courtesy, but little more" (McKnight, 1972).

Meanwhile, Trust Territory conservation laws, which the old people welcomed, were being widely ignored. Dynamite was used to catch fish, thereby destroying reef habitat on which the fish depended for food and shelter. Dugongs, which were totally protected and on the Endangered Species List after being hunted almost to extinction, were still killed surreptitiously. Undersized and out-of-season turtles and their eggs were openly harvested. Small mesh nets captured very young fish.

Trust Territory conservation personnel had little incentive to enforce the few conservation laws on the books. Their lives were sometimes threatened when they did. The chief conservationist had a spear thrown at him one night through his dining room window. Moreover when violators were caught they could depend on very lenient treatment if they had a relative with influence in the government—and almost everyone did; the Palauan extended family extends very far indeed. The chiefs had no legal right to enforce Trust Territory conservation laws and their informal attempts to support them[17] were not always respected. One chief told me that when he tried to reprimand a man for taking an undersized turtle the man replied that the laws governing the taking of turtles were Trust Territory laws and none of his concern.

Those few traditional conservation customs that had survived, such as reef and lagoon tenure, were taken more seriously by potential transgressors. If one was caught breaking a traditional law, punishment was fairly certain and often rather severe by contemporary American standards. "I'm not scared of the court, but I'm scared of the Palauan custom," was a phrase I heard often. "When you are punished Palauan style you are really punished!" But most traditional conservation laws lived only in the memories of a few old men.

Ngiraklang was one of a number of chiefs I talked to who deplored the decline of the conservation ethic. One evening I asked him what he would like to do to rescue Palau's deteriorating environment. "Nothing," he said rather grimly. His answer perplexed me. So a few days later I asked the question again, phrasing it

17. Not *all* the chiefs were concerned. A few helped throw the dynamite and dig the turtle eggs.

differently. "Suppose you were king of all Palau and could make and enforce any laws you wanted to. Would you make any changes to help protect Palau's natural resources?" This time my question elicited a long and thoughtful answer. After he had finished I asked why he had been so negative when I had brought up the subject earlier. "Today we were imagining, and I was the king of Palau," he said. "If there were such a king, he could do something. I can do nothing. I know for I have tried."

He went on to describe how for a number of years he had been a member of the Palau legislature. He had repeatedly introduced conservation legislation but it was never brought to the floor. Every year, for example, he tried to establish laws to control hunting of the diminishing Micronesian pigeon. Ever since guns replaced blow-guns among Palauan hunters, pigeon populations had been shrinking. But the younger men who dominated the legislature were not interested in conservation; either it interfered with their pleasure or with their business. So Ngiraklang finally gave up.

A Renaissance, Perhaps

In the mid-1970s a new generation of Palauans began to make themselves heard. The environmental awareness that mushroomed in the West in the late 1960s was spreading to Micronesia. "We have an entirely different point of view than most of our legislators," one Palauan in his late twenties told me. He continued:

> They were trained to think of our traditions and culture as backward and to think of progress just in terms of money. But we who have recently returned from schooling abroad have learned of the consequences of unplanned economic development, and have begun to realize that our traditions and our culture retain considerable value. That does not mean that we want to shut out the rest of the world. That is impossible. But if we accept technological progress uncritically it could mean the end of our culture and the destruction of the environment in our small islands.

A new awareness was testing the Palauan "conviction that new ways do not mean the dissolution of Palau Society" (Useem, 1955, p. 132).

Ecology became part of the high school curriculum in the 1970s. The results soon became visible. In 1976 teachers and students at Koror High School sponsored a panel discussion on whether Palauans should support plans to develop a giant international oil storage and transfer depot amidst Palau's reefs. A prominent Palauan businessman-politician, who strongly favored the superport but who, it appeared, wanted to seem impartial, addressed the students. If they wanted to be fishermen like their fathers, he said, then he would be against the superport, for it would interfere with fishing. But if they did not want to fish for a living then he would favor the port because it would provide them with jobs. He knew that few students aspired to become fishermen, but he underestimated the depth of their environmental concern. They voiced emphatic and almost unanimous disapproval of the superport.

The superport issue brought forth a novel realignment of social and political allegiances. Old and young were at first unable to form an effective coalition against their middle-aged political leaders; they possessed little understanding of one another and found it hard to communicate. But a leader emerged who represented both traditional hereditary authority and Palau's young adults. The position of high chief of Koror, one of the two highest traditional positions in the islands, fell, unconventionally, to a man still in his twenties. Under his leadership the Save Palau Organization was created to oppose the backers of the superport. A bicameral legislature also came into being. The House of Elected Representatives was joined by the House of Chiefs who now had equal voting rights—a kind of House of Lords to counterbalance the House of Commons.

Intense interdistrict rivalry and strong class and clan affiliations had traditionally precluded any well-developed sense of common identity among Palauans. But the prospect of the superport that, as one observer noted, would have an impact on Palau greater than that of World War II, produced a growing willingness to step across hereditary and regional boundaries to take sides on the issue. A broader, more informed perspective had begun to influence Palauan attitudes toward their heritage. At this writing the future of the superport and Palau's reefs remain undecided.

C H A P T E R

IMPROVING REEF AND LAGOON FISHERIES

Never advise fishermen. If your advice turns out wrong you get blamed. If it turns out right it is they who are the clever ones. It is only from time to time I make, perhaps, a little suggestion.
—F. D. Ommaney

As responsibility for the control of fishing slipped through the hands of traditional island leaders it landed at the feet of government administrations. Island fisheries managers, usually Westerners, have been struggling with the resulting problems for decades, generally with disappointing results. I can think of no inhabited island in Oceania where it can be said that seafood is as plentiful as it was before Western contact. But islands are legion where, concomitant with the decline of local tradition and the growth of human populations, fishing is deteriorating seriously.

With the benefit of hindsight we can now see that what has been needed is not the replacement of tradition with Western science, but rather a wedding of the two. Traditional knowledge must be sought to complement scientists' meagre store of relevant biological information. The effort should not stop there, however, if the union is to be really fruitful. Political, economic, cultural, and social dimensions of a fishery severely restrict the effectiveness of management programs based solely on the biology of stocks. One must study fishermen as well as fish. Understanding fishermen's customary patterns of resource allocation, for example, will help define the context in which biological information may best be employed in managing a fishery.

Reef and Lagoon Tenure

Although fishing ground tenure systems facilitate the conservation of fish stocks, they cannot, without additional restraints, prevent depletion if the needs of the population exceed the maximum sustainable yield of the resource. Palau has no immediate problem here; the ratio of fish stocks to people is high. But export markets may exert pressures that are ecologically equivalent to those of overpopulation; someone else's population may place excessive demands on one's resources. Short-term profits from an export market may loom larger in the minds of those who own the fishing grounds than the long-term disadvantages of overfishing to achieve that profit. The fishermen of Ngeremlengui are a case in point.

But it is important to remember that fishermen in a number of other municipalities have not seriously overharvested their fishing stocks, despite the lure of the market. If Palau's fishing grounds were thrown open to all, overfishing would become chronic and widespread. Aggressive fishermen would fish down accessible stocks to a point where fishing would become unproductive and unprofitable in all accessible areas. The same phenomenon has occurred countless times in Western countries in the absence of limited entry.

There has been a temptation on the part of some island officials, including some of Palau's, to weaken or invalidate traditional marine tenure systems because of certain problems they pose to commercial fisheries development. But such systems need not be perfect to be preferable to unlimited entry. Moreover there are alternative solutions to these problems.

One widespread difficulty concerns the use of tenured fishing grounds by outsiders to obtain bait fish. For example, the crew of a Korean tuna boat licensed to operate in Palauan waters relied for bait on schools of anchovies and herring netted in tenured lagoon waters. Whenever they sought bait fish they had to land and ask permission to fish. The formal presentation and granting of such requests takes considerable time and ceremony in Palau, as elsewhere in the Pacific. The tuna fishermen were so hampered by this problem that they finally left, and valuable resources were left unharvested as a consequence.

Foreign tuna fishermen also left Fiji (Lindsey, 1972) and bait fish areas were closed off by local authorities in New Britain (Kent, 1979) because of similar difficulties. But in two other island groups recent government action appears to have substantially reduced this problem. In the Solomon Islands arrangements have been worked out so that standard payments are made to villagers whenever bait is caught in their waters (Kent, 1979). In Papua New Guinea, local government councils now receive a percentage of the profits arising from the sale of tuna caught by fishermen using bait from their waters (Kent, 1979).

Local fishermen may sometimes be reluctant to invest in the expensive nets necessary to catch schools of migrating fish, such as mackerel, when they cannot pursue them beyond the boundaries of their own fishing grounds. An important source of food and revenue may thus be lost. But if village representatives were encouraged to work out cooperative arrangements, then a single net, jointly purchased at small expense to each individual involved, might be used cooperatively to harvest the school, with portions of the catch or equivalent cash payments going to each village.

Why abandon reef and lagoon tenure systems because of problems such as these when they may be resolved profitably with the help of a sympathetic government mediator? Legislation that nullifies marine tenure laws will, moreover, constrain the former owners from policing these resources—something they do voluntarily if their rights are secure. Destroying a tenure system thus increases a government's regulatory responsibilities[1] and places a heavy additional burden on its typically understaffed and underfunded fisheries department. The government thus disposes of a service it gets for free and assumes new responsibilities it is ill-equipped to handle. Once destroyed such systems may prove difficult to resurrect (Johannes, 1978a).

This is not to say that tenure systems should never be discarded under any circumstances. In Koror Municipality, the seat of Palau's district center, reef tenure rights are no longer observed. Here both migrants from outer villages and foreigners converge, overwhelming the traditional order with new economic, political, and social pressures. Immigrants from other districts greatly outnumber traditional residents in Koror and it is close to impossible to determine (or even define) which fishermen are true residents. In addition Koror's fishing grounds are more extensive than those of most other municipalities and filled with islands among which fishermen can often do what they like, free from surveillance. Enforcement of tenure rights here is thus impractical.

---

1. Maine lobster fishermen have staked out territories, contrary to law, and defended them much as the fishermen of Oceania have done for centuries. The results of their actions, described by Acheson (1972), demonstrate the value to fisheries management personnel of marine fishing tenure.

> When one considers that the entire coast of Maine is patrolled by only a handful of Sea and Shore Fisheries wardens, it is amazing that there is so little trouble. The traditional territorial concepts go a long way in maintaining relative peace.
> The state would do well, when it prepares new legislation on lobstering, to take into consideration the lobstermen's unwritten rules of territoriality.

The destruction of fishing tenure in the vicinity of district centers may be inevitable in Oceania. It has already occurred in many places. It is perhaps even desirable insofar as some portions of the coastline often need be set aside for receiving domestic and industrial wastes on one hand and visitors seeking recreational fishing on the other. Both are often incompatible with good fishing. But if they are confined largely to waters around district centers, local deterioration in fishing conditions is perhaps not too large a price to pay.

It might be thought that fishing ground tenure in Oceania has an unpromising future because of the decline of the power of the chiefs and the weakening of clan and family authority as westernization proceeds. But a different form of authority exists in Oceania with which the responsibility for regulating fishing ground tenure might rest. This is the fisheries cooperative. The transfer of fishing ground tenure from traditional owners to fishing cooperatives was accomplished several decades ago by the governments of Japan and Korea. Fishing cooperatives are gradually taking over the management of some fishing grounds in Palau.

Managing Reef Fish Stocks

The Palauan appetite for imported goods has grown much faster than the economic base necessary to support it. Imports exceed exports by a margin of four to one. Even if the reef fishery were exploited efficiently it would not provide sufficient income to offset this imbalance. But it could make a much greater contribution than it makes at present.

Palau has 2,130 square kilometers of reef, lagoon, and mangrove (Palau Trust Territory Land Office). The average annual yield from intensive reef and lagoon fin-fisheries lies between one and five tons per square kilometer (Munro 1974a; Stevenson and Marshall, 1974). Palau therefore appears to have a potential annual harvest of somewhere between 2,000 and 11,000 tons of reef fish per year. In addition there are significant stocks of crustaceans and molluscs.

Because prices obtained for exported seafood are set by external market forces[2] Palauans are not in a good position to improve their lot by demanding higher export prices. Palauans could, however, increase their share of their own market substantially. By 1974 Palau was importing almost one million pounds of meat and fish annually. This contributed significantly to the trade deficit and to a loss of potential revenue by local fishermen. A conservative estimate of the

2. Guam, the major importer of Palauan fish, also imports reef fish from the Philippines, frozen fish from the North Atlantic, and canned fish from various European and Pacific countries.

potential sustained yield of reef fish from Palauan waters, 2,000 tons, is equivalent to somewhat more than Palau's estimated total consumption of animal protein (see p. 71), and could cancel out an outflow of cash for canned meat and fish amounting to about $500,000 annually, or about $38 dollars per capita. The Palauan government might therefore consider imposing a protective tariff on imported fish and meat to help diminish the loss of foreign exchange and recapture this market. (This has been done elsewhere in Oceania. For example, an import duty on canned fish of 37.5 percent has been levied in the British Solomon Islands, and 34 percent in Western Samoa.)

The largest competitor with reef fish on the Palau market is Japanese canned mackerel. The economics of the mackerel fishery are changing. Much of the harvest is taken in what in the past have been international waters. But with a growing number of nations declaring 200-mile fishing limits, the Japanese are losing control of some of their mackerel fisheries and having to pay neighboring countries for the right to exploit others. This plus rapidly rising energy costs may put canned mackerel out of reach of many islanders within the next few years. And this may help stimulate Palauans to use their own fisheries resources more effectively.

Underharvested Stocks

Exploiting the potential of Palau's reef fishery more fully involves not only managing it more effectively in overexploited areas around Koror, Ngeremlengui, and Peleliu but also opening it up in underharvested areas. (The latter include more than 400 square kilometers of reef and lagoon north of Babeldaob and most of the outer reef slopes stretching for more than 300 kilometers around the archipelago.) Fishermen complain, however, that the cost of gasoline has been going up much faster than the prices they obtain for their fish. Palau's less accessible fish stocks have thus been growing functionally more remote.

If fishermen were reasonably certain of large catches, it would be worth their while to travel to remoter, previously underexploited areas. Fishermen are unfamiliar with spawning locations of fishes in areas where they do not fish. An exploratory fishing program on Palau's remoter reef areas might lead to the discovery of such locations. For example, searching the edges of the west entrance to Kossol Passage and the reef pass about five kilometers south of it around new and full moons (particularly in the spring months when spawning activities reach a peak) might result in the discovery of commercially attractive aggregations of groupers, jacks, snappers, or emperors.

Depleted Stocks

In temperate zone fisheries, mathematical models are used to reach decisions concerning the regulation of fishing pressure. According to Dickie (1962, p. 112) the information necessary to construct one of these models "in its most simplified form [consists] only [of] accurate information on total catch, catch per unit effort and the rate of fishing over a period of years in which there were rather large differences in the amount of fishing done." A typical Pacific island reef fishery involves so many stocks, so many harvest methods, so many fishermen per unit of catch, and so many distribution channels that the effort involved in getting such information and applying it effectively to the management of the fishery would be staggering. The costs would far outweigh the benefits. Conventional fishery models are as useful in these fisheries as sheet music in a frog chorus. (I stress this not because reef fisheries managers do not realize it—most of them do—but because fisheries experts who are trained in the temperate zone but act as consultants in the tropics sometimes do not.)

The government manager of a reef and lagoon fishery such as Palau's has to be content with less than ideal objectives. In the absence of adequate data to determine optimum harvest strategies he must simply try to help steer the fishery on a reasonably steady course between serious depletion and gross underutilization. Inevitably his decisions must rest in no small measure on personal judgment, otherwise he would be unable to reach any decisions at all.

In Palau species whose stocks appear to be particularly heavily depleted include those shown in table 4. Any form of management employed to rebuild these stocks must be inexpensive, involve only a few management personnel, and be simple technically; Pacific island governments like Palau's cannot justify spending large amounts of money to manage their small fisheries. Major catches of the species in category A, table 4, are made while they are in their spawning aggregations. The timing and locations of these spawning aggregations of most of these species can be predicted accurately (see chapter 3 and appendix A). A simple means of reducing the fishing pressure on them, therefore, would be to prohibit their sale during all or a portion of their spawning period. In Koror only a single individual would be needed to enforce such laws, and this activity need occupy only a limited portion of his time because of the limited periods during which spawning occurs (see chapter 3).

If only the sale of a species were prohibited, fishing for it for communal use would still occur. In addition some illegal sales would undoubtedly be made in outer villages where surveillance and

TABLE 4.   Fish Species Whose Populations Have Declined Markedly in Apparent Response to Fishing Pressure in Palau According to Fishermen.

A.   Species harvested particularly heavily while in spawning aggregations.

1.   *Herklotsichthyes punctata—mekebud*

2.   *Crenimugil crenilabus—kelat*

3.   *Plectropomus leopardus—mokas*

4.   *Epinephelus merra—mirorch*

5.   *Lethrinus miniatus—mlangmud*

6.   *Siganus canaliculatus—meas*

7.   *Siganus lineatus—klsbuul*

B.   Species harvested particularly heavily by spearfishermen at night.

1.   *Bolbometopon muricatus—kemedukl*

2.   *Cheilinus undulatus—mamel*

enforcement of the law might be impractical. This would be acceptable in some cases. But where more rigorous control of very heavily depleted stocks is desirable, a ban on the capture of the species during spawning periods would be necessary. This would exert a greater sparing effect on stocks, but would require more enforcement personnel. Enforcement would only be required, however, during the limited periods of the year and lunar month and in the limited areas on the reef in which major spawning aggregations are exploited. For example, enforcement of a ban on the harvesting of the badly depleted spawning runs of *meas, Siganus canaliculatus,* would require surveillance at only a few specific locations along the reef edge and only for several days, beginning three days after the new moon, in April and May (see p. 182 for a description of the spawning runs of this species).[3]

3. When *meas* are heavily depleted over a period of years on one spawning ground this does not seem to affect the numbers that spawn on other less heavily fished spawning grounds according to Palauans. Thus different spawning aggregations within Palau lagoon may represent different populations. This is surprising, given the offshore pelagic juvenile stage in their life history. But if it is indeed the case, attempts to replenish *meas* runs could be focussed independently on individual spawning aggregations. The extent to which different spawning aggregations of this and other species of reef food fish represent different populations might be determined by genetic and tag and recapture studies. The results would have important implications for reef fisheries management.

Mekebud, Herklotsichthys punctata, are harvested while in their spawning aggregations, but they are not readily amenable to the management procedures described in the foregoing. Their aggregations sometimes occur in tidal creeks where surveillance of fishermen is difficult, and they are not generally sold in Koror where sales could be regulated. They are used mainly in the villages as food and bait. Populations of these fish have apparently fallen drastically in the past twenty years, probably due to overfishing. Kramer (1929, p. 112) mentions "great schools" of *mekebud* in the lagoon. Ngiraklang remembered getting two and one-half fifty-five-gallon drums full of *mekebud* with a single can of dynamite while fishing off the Ngeremlengui dock for the Japanese in the 1930s. Today the entire run at Ngeremlengui amounts to less than one drum full of fish and in some years fail to materialize at all. Similar declines are reported from the area of the shipyard near Koror and at other spawning locations in the Rock Islands and around Babeldaob. An attempt to rebuild these once valuable stocks seems particularly desirable[4] because of their potential value as bait for commercial tuna fishing (Baldwin, 1977; Kearney and Hallier, 1978). But public education would have to be coupled with legislation, as laws controlling the harvest of this species would be largely unenforceable by means of compulsion and would require voluntary compliance to be effective.

Direct surveillance of spawning aggregations of these and other species by fisheries officers could provide rough indications of population size. These observations could be used to monitor stocks and aid in deciding whether regulation of the catch is needed and how stringent it should be. Conventional studies of recruitment growth rates, catch per unit effort, and so on would all contribute to a refinement of such an approach to management, but in many instances the costs would outweigh the benefits.[5]

Another form of regulation may be desirable for species that are overharvested as a result of nighttime underwater spearfishing. Catches of at least two of these species have declined markedly in

4. As with *meas,* spawning aggregations of *mekebud* have been observed to diminish greatly in some heavily fished locations without those in other less intensively fished areas decreasing concomitantly. This suggests that different spawning aggregations may be isolated genetically, and that attempts to rebuild *mekebud* populations might be tried out initially on a single aggregation.

5. Such studies have been carried out on reef food fish in Jamaica by John Munro and his colleagues. Here, as well as in some other parts of the Caribbean, reef fishing is carried out largely with portable fish traps. Reef fish in spawning aggregations tend not to enter traps according to fishermen (see also Morrill, 1967, p. 405). The scheme I am proposing here would therefore be ineffective in a trap fishery. Under such conditions the trap mesh-size regulation recommended by Munro (1974a) on the basis of the Jamaica study clearly constitutes a more appropriate form of management (see also Stevenson, 1977).

Palau after the introduction of underwater diving lights. (As mentioned earlier, there is some evidence, based on comments of fishermen and fisheries officers in various parts of Micronesia, that some of these species tend to rest in deeper water after exposure to a few years of nighttime spearfishing. Here they are less accessible to divers and the practical consequences are almost the same as if the species is being overfished.)

Spearfishing at night has been banned on certain other Pacific islands as a consequence of these problems (Johannes, 1978a). It would be impractical to prohibit night spearfishing in Palau, however. Effective surveillance would be impossible in most areas. Moreover spearfishing is engaged in to a large extent by younger men who are not skilled in the use of other techniques. It is an efficient and widespread method of obtaining food for one's family that involves a minimum of capital outlay. And it enables fishermen to capture species that are otherwise rather inaccessible. If some form of control seems desirable, prohibiting the commercial sale of target species might be considered. This would decrease the fishing pressure on these particular species without interfering unduly with an important subsistence fishery.

When fishermen do not understand the purposes of fishing regulations or perceive them as being imposed arbitrarily by outsiders they are not liable to look on them with favor or obey them voluntarily. There is a major advantage to regulations that have a precedent based on local custom; they are liable to be viewed with relative sympathy by fishermen. A ban on the harvesting of spawning aggregations, for example, was once practiced traditionally in parts of Palau (p. 67) as well as other parts of Oceania (Johannes, 1978a). Palauan fishermen themselves have recently initiated legislation that prohibits fishing in Ngerumekaol Channel during the months of May, June, and July when several species of groupers aggregate there to spawn.

Similar approaches to fisheries management might be tried in other coral reef fisheries.[6] This would require studies of the timing, location, and exploitation of spawning aggregations as well as other aspects of the behavior of fish and fishermen—studies to which fishermen might contribute greatly.

6. The general aspects of this approach to management are discussed in more detail elsewhere (Johannes, 1980).

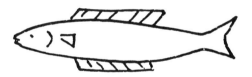

C H A P T E R

FISHING IN THE SOUTH WEST ISLANDS 7

R. E. Johannes and P. W. Black

In Palau until recently there has been little incentive to fish beyond the extensive reefs and wide lagoon, for plenty of fish could be caught within their shallow, sheltered confines. But several tiny islands to the south of Palau have not been so favored by nature. Their inhabitants have of necessity fished routinely for centuries over the deeper, rougher waters of the outer reef slope and the ocean beyond. I wondered how the skills and perceptions of such fishermen would differ from those of Palauans. The opportunity to find out presented itself conveniently.

On the edge of Malakal Harbor near Koror were a cluster of small shacks of plywood and corrugated metal that housed several hospitable families from one of these islands, Tobi. They had moved to Palau in order to avail their children of the schools there. Among them were some of Tobi's best fishermen. One of them, Patris Tachemaremacho, became my chief Tobian teacher-interpreter. Patris's knowledge of Tobian fishing was considerable, but the ultimate authority on most such matters was his father, Patricio. (When it came to the spelling of Tobian fish names, however, everyone deferred to Patris's wife, Marina).

Now in his eighties, Patricio was still mentally quick. He had been one of the most respected teachers of fishing and canoe handling on Tobi. (He had also been one of the main informants of Peter Black, who collaborated with me on this chapter. Black had lived on Tobi, first as a Peace Corps worker, later as an anthropologist.[1]) If a question came up during our interviews about which Patricio was unsure, one of Patris's numerous handsome children was sent to find two other expert Tobian fishermen, Kalisto Elias and Zacharias Saimer.

Tobi and its sister islands, Sonsorol and Pulo Anna, are among the smallest and most isolated of the inhabited islands of Micronesia. They lie 200 to 300 miles south of Palau in the southwestern corner of Micronesia. Although separated from each other by 80 to 175 miles of open ocean, their inhabitants are united by a common language and culture.

The significance of these islands in the global scheme of things is such that no one in the West has ever bothered to give them a collective name. They are known today simply as the South West Islands. Many people in the Trust Territory Government are not even aware of their existence. The language and culture of the inhabitants are quite different from those of Palau, but for geopolitical convenience they have been designated as part of the district of Palau. Furthest from Palau, Tobi is the least visited and the most traditional of these islands. The only regular contact with the outside world is provided by a small government supply vessel that comes from Palau several times a year, stopping for a few hours to offload sugar, rice, and cigarettes and pick up copra.

Tobi covers an area of 0.23 square miles. Phosphate deposits render the soil more fertile than it is on many low tropical islands. But in times past the population pressure on these terrestrial resources has been very great. Today there are only about sixty people on Tobi, but starvation due to overpopulation was apparently once not uncommon (e.g., Eilers, 1936). Tobians say there were times when resources were subdivided so finely, due to overpopulation, that different individuals owned different fronds on the same palm tree. In the early 1900s the population reportedly stood at 986 (Eilers, 1936). If accurate this figure yields a remarkable population density of more than 4,000 per square mile.

---

1. Significant portions of this chapter are derived from field notes and an unpublished manuscript on Tobian fishing methods written by Black (1968). In his thesis on Tobian culture he wrote of the "possibilities which these thoughtful, intelligent and verbal [Tobian] men offered for the investigation of [native] fishing." He also noted that "their eagerness to record for the future what they felt to be a passing set of skills was impressive" (Black, 1977, p. 30).

On Tobi, unlike Palau, there are no forests, streams, lakes, or mangrove swamps. Thus there are no jungle fowl, freshwater fish and shrimp, fruitbats, or mangrove crabs. (A single stray crocodile reached the island a few years ago and was speared and eaten.) A narrow fringing reef provides the only significant source of animal protein other than the open ocean. Such shallow coral reef communities may, when intensely harvested, yield as much as twelve metric tons of fish and shellfish per square kilometer (Hill, 1978).[2] Because Tobi's reef occupies an area of about one and a quarter kilometers, its maximum annual yield of seafood would probably amount to no more than about fifteen metric tons. Shared by 986 people, this would have allowed each an average of only about thirty-six grams of flesh or about ten grams (about one-third ounce) of protein per person.[3] The reef therefore could not have been the sole source of animal protein. (Curiously Tobians did not eat a variety of easily harvested reef invertebrates [e.g., sea cucumbers, sea anemones, sipunculids, starfish] eaten by Palauans and other Pacific islanders.) It can thus be seen why Tobians have depended heavily on the waters beyond the reef for their catch. Of forty-one Tobian fishing methods listed by Black (1968), thirty-three involved fishing beyond the outer reef crest.

So steeply does the bottom fall away as a canoe pulls past the breakers on Tobi's reef that within seconds only blue water can be seen beneath. A trip of a few hundred meters takes a fisherman from his house to the open-ocean fishing grounds where he seeks tuna, dolphinfish, rainbow runners, wahoo, marlin, sailfish, barracuda, flying fish, and sharks.

Until recently the fishermen of Tobi also sailed periodically to Helen Reef, a tiny uninhabited atoll thirty-nine miles to the east, in order to get turtles and giant clams.[4] According to informants these clams (*Tridacna*) were once common on Tobi's reefs but were virtually exterminated by overharvesting. (A single large giant clam was discovered on Tobi in forty feet of water in 1968.)

---

2. This value is much higher than other published yields calculated for reef fisheries (e.g., Stevenson and Marshall, 1974). This is because the other estimates relate only to commerical catches, whereas Hill (1978) measured yields on Samoan reefs that were heavily gleaned on a subsistence basis for small invertebrates and fish as well as the larger commerical type specimens—in other words, reefs harvested with similar intensity to Tobi's.

3. This figure is calculated assuming fresh reef fish average 27 percent protein (Murai et al., 1958).

4. The Tobian name for this atoll is *Hocharihie*, "Island of the Giant Clams." Sadly this lovely place, containing the most beautiful reefs I have ever seen, has been one of

Learning to Fish

Traditionally Tobian boys went through long and rigorous training before mastering all the fishing techniques to which they could legitimately aspire, generally reaching middle age before having a chance of being considered a *marusetih* or master fisherman. At seven or eight years of age they would begin their fishing careers. Fishhooks were too precious, prior to the introduction of metal, to be entrusted to neophytes. Consequently a young boy's fishing gear consisted only of a very fine line. A loop was tied on the end, in the middle of which a piece of hermit crab was suspended as bait. Gathering on the reef flat at low tide the boys used these simple devices to noose small fish in tidepools. A small unidentified jack was especially sought after; the rough lateral keel on its tail helped prevent the noose from slipping off it. The boys were restricted to this kind of fishing for three or four years. Fishing at such close range to the fish, they learned much about the behavior of different species.

In the meantime they were taught how to make fishhooks out of shell or bone. Only after they could make them well—a demanding operation (see Chapter 9)—were they allowed to use them. Their training continued in a series of steps involving the use of poles, hooks, and lines on foot in shallow reef waters. The final stage in this period, when the boys had reached late adolescence, involved learning how to cast out beyond the reef crest to catch larger fish living on the outer reef slope. This was the first time the boys engaged in a fishing technique routinely used by adult fishermen.

This traditional training schedule exists today in attenuated form. Young boys are encouraged to fish on the reef for small fish but the procedure is less rigid than it used to be. Baited hooks are used because metal hooks are now cheap and expendable. Otherwise boys go through the same general progression as their ancestors, but in a much shorter time. As on most tropical Pacific islands, there were numerous onerous fishing taboos and protocols, often of a religious

---

the subjects of systematic pillage by Okinawan, Taiwanese, and Korean fishermen. These men have ravaged reefs from the Marshall Islands through western Micronesia and all the way to Australia's Great Barrier Reef in their search for giant clams. In addition to decimating the atoll's giant clam populations (Bryan and McConnell, 1976), a party of Korean fishermen recently devastated rookeries of several species of sea birds in their quest for food and cut down most of the islet's palm trees, rather than climbing them, to get the coconuts.

Helen Reef and Merir, another uninhabited island in the South West Islands, are the most important green turtle nesting sites anywhere within U.S. jurisdiction (Pritchard, 1977). Unfortunately a small number of Palauans habitually harvest turtles illegally during visits of the government supply vessel to Helen Reef. South West Islanders resent these intrusions—illegal according to both traditional and modern laws but cannot stop them; the Palauan government has simply ignored complaints concerning these activities.

nature, that had to be learned by apprentice fishermen. But with the introduction of Christianity many of these practices ceased (Black, 1977). Today boys are sometimes allowed to fish beyond the reef as soon as they are physically capable of doing so, due to a shortage of manpower. Depending on their fathers' judgment of their physical maturity and strength, boys from thirteen to sixteen are now crewing on the large canoes and trolling for tuna or netting flying fish. After a year or two of instruction by older male relatives the boys start going out with friends and brothers.

In theory no Tobian can master all traditional fishing techniques, for some are considered private property. (Virtually the only things that are owned outright among the communal Tobians are nonmaterial. These include not only fishing techniques but also certain songs and medicines.) As a general rule the more difficult a technique, the fewer people that own it. A boy gaining his initial access to a one-man canoe is told by his father which of the techniques he may legitimately use. Once having mastered these techniques he is drawn to those belonging to others and he proceeds to try to "steal" them. (This is the Tobian term and refers to acquisition of a technique by covert observation.) Arguments concerning this practice are frequent. Serious confrontations are avoided, however, by directing accusations and retorts back and forth from a distance and over an extended period of time through a chain of intermediaries.

Such ownership practices have eroded among South West Islanders in Koror; many of their techniques are inapplicable in the lagoon fishing areas to which they have easy access. Only once during my interviews in Koror did a man explain politely that he could not answer my question because the relevant knowledge was a family secret.

Traditional Marine Conservation

Secrecy probably functions as a conservation measure; if the knowledge needed to exploit a particular area or species is restricted, the likelihood of overexploitation is lessened (e.g., Forman, 1967). (Conversely the "stealing" of a method helps reduce the risk of it being lost if its legitimate owner dies without heirs.) A variety of other South West Island customs also functioned to conserve certain species. Some of these practices were imbedded in ritual and it is impossible today to determine whether or not they were originally intended as conservation measures.

Octopus, for example, could not be eaten on Tobi except by the elderly. Octopus make excellent bait but are not abundant around Tobi. Was this restriction set forth originally in order to maintain the bait supply? Today Tobians no longer remember. A variety of other species were taboo to part of the population or during part of the year. Still other edible species were forbidden to all at all times. Such complete prohibitions *appear* not to have been erected as conservation measures as their effect seems functionally equivalent to exterminating the species in question. However we cannot assume that when survival was at stake Tobians did not set aside such taboos.

Some South West Island customs were clearly designed with practical conservation in mind. As described earlier, seabirds help fishermen find fish. In a conscious effort to conserve the populations of these birds it was forbidden on Sonsorol to eat them except during the nesting season when they were very abundant. The use of fish poisons was also restricted to certain special occasions on Tobi and Sonsorol because it killed too many juvenile fish, thus reducing the future supply.

Although turtles were never abundant around Tobi within living memory (see also Holden, 1836), their numbers seem to have decreased even further in recent years. Several years ago it was decided at a meeting that turtle eggs (a great delicacy) would no longer be eaten, so there would be more turtles to eat in the future. Anyone who violated the new law would be fined. A person finding a nest of turtle eggs reported it to the magistrate who immediately fenced the site to keep the hatchlings safe from cats. When the eggs hatched (the time can be predicted to within a day or two) the hatchlings were not allowed to make their dangerous trek across the beach and reef to the open sea. To the apparent frustration of the many seabirds that gathered at the first sign of hatching, the hatchlings were gathered up. They were kept in a large bucket and fed finely chopped fish. When they were judged big enough to have a good chance of surviving they were ferried by canoe out to the open sea and released.[5] (Unfortunately, a new crop of teenage boys not in on the original decision began eating all the eggs they could find recently.) A similar conservation measure was introduced at about the same time on Sonsorol.

FISHING METHODS

Several dozen different fishing methods are used in the South West Islands. Trolling for large pelagic fish and dipnetting flying fish by torchlight are by far the most important methods in terms of

5. The extent to which turtles depend on their trip across the beach and reef in order to "imprint" on their birthplace and find it again at egg-laying time is unknown. If this trip is an important part of the imprinting process then these efforts at conserving turtles may, in fact, deplete them even further.

yield. Only when they are unproductive is much time spent on other types of fishing. Here, briefly, are described a few of the more important or interesting methods.

Trolling

Of all activities connected with the sea, Tobian men get most pleasure out of trolling from sailing canoes. When the tuna are there the pace of island life quickens. The men put out in their canoes morning and evening, beating back and forth in the wind, waiting for the shock of a big fish as it hits the feathered trolling lure. Working their way gradually offshore, sometimes beyond sight of the island, they search for the feeding birds or the small fish breaking to surface that signal the presence of bigger fish below.

Handling sailing lines and fishing lines at the same time in stiff winds and rough water is no easy matter in their small canoes. This is particularly true when coming about, during which the whole sail rig must be lifted over the fishing lines, carried along a lurching deck no more than eighteen inches wide, and placed at the opposite end of the canoe. To add to the challenge, sharks sometimes crash into the side of the canoe while chasing a hooked tuna, which in turn may put heavy strain on the line when diving to escape its pursuers. It is not surprising under such conditions that canoes sometimes capsize. The catch is the only thing in the canoe that does not float, so one man is delegated to dive after the fish and bundle them up with sail, poles and paddles, while the other man rights and bails out the canoe.

Racing breaks the monotony when fishing is slow. A passage from Peter Black's diary:

> Lawrence came out late, close between the island and our canoe as we rounded the southern tip of the island, . . . the area of choppiest seas and most gusting wind. Giant waves breaking on the reef, the sun breaking through the thunderheads on the eastern horizon, Lawrence's sail sparkling in the brand new sunlight. Lawrence was going flat out, challenging us to a race, hauling full strength on his sail, with the outrigger and its platform pointing up at the sky and Lawrence squatting on the tilting and lurching canoe, laughing and singing and dancing.

When the fish are biting but the wind stops, the canoes are paddled back and forth in order to troll. This is hard work and the catch is generally not as good as when the wind is up.

Most fishermen go in pairs. The only people who sail the big sailing canoes alone are those who have dependent youngsters at home whom they can summon to help bring in the canoe. This is accomplished by means of a unique call, made by scraping a piece of the canoe's plywood deck with considerable pressure across its gunwales. The entire canoe becomes a sounding board that emits a loud, resonant groan. The sound is distinctive for each canoe, and recognized by the helpers on land.

Torch Fishing for Flying Fish

On spring nights flying fish are attracted to canoes by the light of palm frond torches and dipnetted from the water. This is done only during the twelve dark nights of the moon; torches are ineffective on bright nights when the moon tends to drown their light. A netman must not eat, drink, or smoke from the time he gets up in the morning until when he returns from the sea. The rest of the crew observes the same restraint after the midday meal.

Well after dark the canoes slip out through the reef channel. The crews are under the direction of taciturn netmen. Only the breakers, the vague outline of the island, and the stars can be seen in the darkness. Each canoe is paddled into position and the torches lit. The scene is instantly transformed as domes of light illuminate each canoe and its paddlers. The images are repeated beneath on a mirror-calm sea. Beyond the domes nothing is visible.

Swimming close to the surface the fish are caught with a downward sweep of the dipnet, taking advantage of their tendency to swim upward when startled. Sharks and needlefish are also often attracted by the light. The dip and heave of the net continues uninterrupted, without interference from sharks, although their shapes often loom close and large.

The needlefish sometimes pose a real problem, however. Somehow stimulated by the torchlight, they periodically erupt from the water and rocket through the air, their slender bodies and long, sharp, bony snouts giving them more than a superficial resemblance to javelins. The fishermen, who traditionally go naked while torch fishing, run the risk of being seriously injured. At least two Sonsorolese fishermen have been killed as a result of wounds from these fish—one by a neck wound, the other by an infection in an abdominal wound.[6] (Eilers [1936] mentions a Tobian medicine made expressly for treatment of needlefish injuries.) Needlefish are particularly abundant around Sonsorol and once they reach numbers judged by the chiefs to be too hazardous, torch fishing is called off for the season.

---

6. This problem is not restricted to the South West Islands. People throughout the tropics have been killed or injured by flying needlefish (e.g., Guppy, 1887; Kayser, 1936; Pratt, 1906; Randall, 1966). In Hawaii in 1977 one person was injured and another killed by needlefish (*Honolulu Advertiser*, September 6, 10, 1977).

Kite Fishing

Fishing with a kite made from a leaf attached to a line with a hookless lure is indigenous to the South West Islands, as it is to certain other areas scattered throughout the tropical western Pacific and Southeast Asia.[7] In the South West Islands the kite is made from a breadfruit leaf. The leaf is dried by passing it over a fire and pressing it flat under a woven sleeping mat. The slender, dried midribs of coconut leaflets are threaded through the breadfruit leaf and tied to one another where they cross to give it rigidity. The line is made from sennet (coconut husk fiber).

In strong winds the kites are reportedly flown as high as 300 feet. The bridle can be moved up and down to adjust for wind strength: the greater the wind speed the smaller the angle between the kite face and the horizontal. When kite fishing while standing on the outer reef edge, a fisherman can alter the direction in which the wind carries his kite by trimming it on one side to make it fly to the left or right. This enables him to get the kite offshore even if the breeze is running parallel to the reef edge or blowing obliquely onshore. When fishing from a canoe the fisherman ties the kite line to his canoe or holds it in his mouth once the kite is up, so that he can paddle the canoe in a zig-zag upwind course. (Children on foot fish from the reef edge using smaller versions of these kites with the line attached to a long rod.)

Dangling from the kite is an extension of the kite line bearing a lure made from the web of an indigenous spider. It serves simultaneously as the kite's tail. Ten to fifteen webs are twisted onto a forked stick to make the lure, which looks like a small gauze pad. Its shape varies depending on the maker. The kite is maneuvered so that the lure skips and splashes along the surface of the water. Needlefish strike at it, apparently mistaking it for a jumping fish. The many small, sharp-pointed teeth in their long jaws get tangled so tightly in the webbing that the fishermen sometime have to cut fish out after they are landed. Catches of half a dozen three- to six-foot needlefish in half an hour are not uncommon when the fishing is good. On rare occasions barracuda and mackerel are also caught in this manner.

On Sonsorol the species of spider that builds the appropriate web does not occur. Attempts by Sonsorolese fishermen to introduce it from Tobi have not been successful. (This is surprising because the fauna, flora, and landscape of the two islands are very similar, and whatever thrives in one might be expected to thrive in the other.) So to make their kite lure, the Sonsorolese use the connective tissue

---

7. Locations where kite fishing has been used, and variations on the general method, have been described by Balfour (1913) and Anell (1955).

lying just beneath the skin of lemon sharks, *Negaprion acutidens*. It has the disadvantage of becoming dry and leathery when stored and must be chewed to render it supple before each use.[8]

A drawback of the spider web lure is that contact with even a small amount of oil makes the fibers slippery enough so that the teeth of the needlefish tend not to hold in them. A fisherman must wash his hands well before touching the lure so that the skin oils on his fingertips do not spoil it. Coconut oil and, these days, hair pomade are also avoided for three days prior to kite fishing. Around Patricio's adopted home in Palau a thin film of oil on the water emanating from nearby boat docks and a copra-processing plant render these lures unusable.

Noosing Sharks

Fishing line made from plant fiber cannot resist for long the teeth of a good-sized shark. Lacking metal leaders, Pacific island fishermen solved this problem in a number of ways. One was to fish with wooden hooks so large that a portion of the shank protruded from a hooked shark's mouth, preventing its teeth from coming in contact with the line. Another method, mentioned in chapter 2, was to noose sharks around the gills, thereby not only keeping the line away from the shark's teeth but also cutting off the shark's breath and thereby shortening its struggles. First, however, the shark had to be attracted close to the canoe.

Centuries before biologists, Pacific islanders discovered and put to use the knowledge that certain types of vibrations will attract sharks (e.g., Hodgson, 1978). (Some vibrations have recently been found by biologists to be even more effective in this regard than the odor of bait [Nelson and Johnson, 1976].) Typically islanders strung together coconut shells and shook them beneath the water to bring sharks close to their boats. South West Islanders used seashells held over the side of the canoe for this purpose, rubbing them together or rattling a string of them rapidly and repeatedly. The fishermen of Pulo Anna were particularly adept at this technique. According to Patris both the frequency and the rhythm of the sounds thus created were important and it took time to acquire proficiency in attracting sharks this way. Recent research suggests why. Not only are closely spaced sounds more effective in attracting sharks (see Chapter 3) but their attractiveness is further enhanced if their rhythm is irregular— that is, if groups of sounds are interspersed with pauses of varying duration (e.g., Myrberg, 1978).

---

8. In the Gilbert Islands I was told that a piece of the dried, pounded intestine of a balloon fish is trailed along the surface by means of pole and line to catch needlefish.

Once lured to the vicinity by these sounds, a shark would be attracted closer to the canoe by raising and lowering in the water a white rock attached to a line. The shark would often be seen to pursue the stone. (The Tobian name for the lemon shark, *Negaprion acutidens*, is *echarivus*, which can be translated as "[bites] white stones"). Once a shark was spotted it was lured still closer to the canoe with a fish trailed in the water. The noose was lowered between the baitfish and the oncoming shark. As the shark swam into the noose, it was pulled tight against its abrasive, nonskid skin.

A skilful variation on this technique was used to catch sharks in deeper water. The bait was suspended in the center of a heavy, stiff noose by means of strings attached to the noose like crosshairs in a rifle sight. The noose was lowered to depths of as much as sixty to eighty fathoms, well out of sight of the fisherman, who determined by the feel of the line when a shark was taking the bait. As the shark mouthed the bait, the strings broke and its head pushed through the noose. But its pectoral and dorsal fins prevented its body from sliding smoothly through. The fisherman monitored this action by the feel of the line and pulled the noose tight at the appropriate moment.

Some sharks swam upward with the bait in their mouths creating slack in the line and reducing the ability of the fisherman to feel what was happening. Consequently the noose was continually raised and lowered while fishing so as to guard against a shark getting away with the bait in this manner before it was detected. Sharks six to eight feet long were caught in this fashion. Larger sharks were usually not sought because they were too strong to be handled safely from a small canoe. Noose fishing is no longer practiced on Tobi.

For sport, tuna were occasionally noosed by a different method. When a feeding school was near the canoe they were attracted closer with chum. A small, flexible noose was lowered and the tuna noosed around the tail with a quick jerk as they inadvertantly swam through the noose. Tuna, like the jacks noosed by children, have a stiff caudal fin and a rough, rigid lateral keel on the tail-base, which prevented the noose from slipping off easily. Needless to say very quick reflexes were needed to fish in this manner.

Log Fishing

The subject of drifting logs is one of great importance to the inhabitants of small oceanic islands in the central west Pacific. The observations of fishermen on the seasonal distribution of these logs and the aggregations of fish beneath them illuminate more than one problem of interest to students of the sea.

For reasons that remain to be discovered, many tropical pelagic fishes tend to gather, often in large schools, under logs and other floating objects. Schools of dolphinfish (mahimahi), jacks, oceanic triggerfish, and most importantly, skipjack tuna will often form under a floating object as small as a single palm frond.[9] Although various small forage fishes also aggregate under these objects, it seems unlikely that their presence is what attracts the larger fish. Tuna schools sighted under single drifting logs in the central west Pacific sometimes weigh more than fifteen tons and consist of several thousand individuals. The few small fish sheltering under the same log could hardly qualify collectively as an appetizer for such aggregations.

Excitement spreads quickly through the village when a drifting log is spotted from the beach on Tobi. Fishermen grab their fishing boxes from beneath the thatched roofs of their canoe houses and launch their canoes. Anyone can fish around a log. But traditionally the first men to reach it gain the right to the log itself. And on such a small, isolated limestone island with a very limited range of in- digenous raw materials, a log, plus the soil and rock carried in its roots, can be a precious commodity. The wood, used as building material and for canoes, was particularly valuable during periods of overpopulation when the demand for wood outstripped the island's supply of suitable trees. The black root-soil is rinsed with rainwater to remove the sea salt and used to enrich the earth in which papaya trees are planted. When mixed with the juice of a tree bark it once also provided a cloth dye.

The hardest objects normally found on limestone islands such as Tobi are relatively soft seashells and coral rock. The harder igneous rocks brought in tree roots were thus once of great value as tools. One form of black rock was used to form the body of a type of tuna lure. Because Tobians had no pottery they had no utensils in which liquids could be heated over a fire. Certain liquid medicines

9. Tobians and fishermen from the eastern Carolines I talked with both noted a relationship between the characteristics of a log and the abundance of fish under it. The older the log and the more gooseneck barnacles and seaweed hanging from it, the more fish it attracts (see also Inoue et al., 1968). Mahogany logs have few fish under them if they have no bark. Clumps of bamboo with withered leaves still attached seem particularly attractive.

There is a saying on Tobi that coaxing fish out from under a log is like coaxing a white man's ship to the island: Plenty of sweet chum (chopped-up fish) must be used in one case and plenty of sweet words in the other. Tobians had an exaggerated reputation for ferocity in the nineteenth century, due to the highly colored descrip- tions of them given by Holden (1836). Thus captains who might normally put in for provisions when sighting an oceanic island tended to steer clear of Tobi (Eilers, 1936). The Tobians, however, in order to obtain metal and tobacco, wanted them to stop and trade. Consequently they often chased passing ships in their canoes, probably inadvertently adding to their reputation for fierceness in the eyes of some sailors who could not tell the difference between entreaties and threats when uttered in the guttural language of Tobi.

were warmed, however, by dropping an igneous rock, heated red-hot in a fire, into a wooden bowl or clamshell containing the medicine. (Calcareous rocks and shells do not hold heat well and tend to flake or shatter when placed in a fire.)

The "season of drifting logs" starts at Tobi in July, with the best months being September through December. This seasonality has two causes. First, it is mainly during the rainy season that trees are washed from river banks in New Guinea and the Philippines and carried downstream to the sea, some of them ultimatel, reaching Tobi. Second, the prevailing ocean currents on which the logs are carried shift back and forth over several degrees of latitude with the seasons.

A similar calendar of log months was described to me by the fishermen of Sonsorol and of some of the outer islands in the eastern Carolines. They often remarked that the number and average size of logs drifting past their islands seem to have decreased considerably in recent years. Perhaps this is the result of accelerated logging in the forests where the logs originate.

This information has potential value for commercial tuna fishermen in the central western Pacific. The area contains the only major underexploited skipjack tuna populations in the Pacific. At present a purse seine fishery is developing in the area and this fishery is largely dependent on floating objects for its catches. Tuna schools can be caught with greater ease in the eastern Pacific because the fish will not dive to escape the encircling net; a shallow and abrupt vertical temperature gradient (thermocline) through which the tuna are reluctant to pass restrict them to shallow water. In the western Pacific, however, this thermocline is deeper and allows the tuna to dive beneath the net to escape.

In the western Pacific, consequently, tuna seiners set their nets mainly around schools associated with floating objects. Once an object with an associated school of tuna is spotted, a radio beacon is attached to it. Its position is then monitored electronically. Only around dawn or dusk is a net set: the object is to make a set when there is just enough light for the fishermen to see what they are doing, but not enough for the fish to see the encircling net and dive in time to escape it.

Tuna seiners in the central western Pacific search almost at random for floating logs. Yet, as described above, the testimony of South West Island fishermen indicate that the logs are not distributed at random. Consequently if the knowledge concerning "log seasons" possessed by native fishermen from small islands throughout the area were gathered and integrated, a picture of the seasonal and geographic variations in log abundance should emerge. This

could lead to significant fuel savings and higher catches for the purse
seiners and hasten the more efficient utilization of these important,
underutilized fish stocks.

Observations of fishermen concerning some of the smaller fish
associated with drifting logs proved interesting. Tobians and marine
biologists alike have noted that the young of a variety of species of
reef fish are often found associated with drifting logs far out to sea.
This is the consequence of many reef fish having pelagic eggs and
larvae. Only after a period of weeks in the water column do the
young take up residence in a reef community. In the meantime floating logs provide some of them with shelter from predators among
their roots and branches.

Ultimately these fish must reach shallow reef areas if they are to
spawn and complete their life cycles. Some species, according to
Patris, will abandon their floating shelter and swim toward Tobi's
reefs when they are still many hundreds of yards away and well out
of sight of them. Such species include tripletails (*Lobotes surinamensis*), triggerfish (*Odonus niger*), and unidentified jacks. (Ngiraklang said he had seen other juvenile reef fish, including rabbitfish
[*Siganus canaliculatus*], emperors [*Lethrinus harak*], and goatfish
[*Mulloidichthys* sp.], abandoning logs off Palau in the same manner.)
Perhaps these fish detect the presence of nearly reefs by means of
olfactory cues; the metabolic activities of the organisms in communities results in the export of biogenic substances downcurrent in the
water that has washed over them (e.g., Johannes, et al., 1972). The
ability to detect reefs at a distance would increase the likelihood of
pelagic juvenile reef fish completing their life cycle.

Fly Fishing Tobi Style

Tobians are adept, like most skilled line fishermen, at determining by the feel of the line what is biting and just what is happening with the bait. Consequently they almost never leave their
lines untended. An exception to this involves a sort of fly fishing,
where the energy of waves, rather than the fisherman's arm, is used
to manipulate the fly. A stick is inserted in a piece of cork made from
driftwood. A few inches of line with a small, white-feathered lure is
tied to the top of the stick. The length of line is adjusted so that the
lure is just below the surface when the cork and stick are floating
upright. A dozen or so of these devices is tossed overboard in the
calm waters just upstream of the island, where wave action is
typically gentle. The small waves tip the cork to and fro and cause
the lure to dance in and out of the water. The action of the lure
attracts flying fish. The fisherman knows he has a fish when one of
his rigs upends.

Netting Triggerfish

The *fen*, which looks like a circular crab net, is used in an unusual fashion to catch triggerfish. Bait is suspended below the center of the mouth of the net by a string tied at the ends to opposite sides of the hoop. The device is lowered from a canoe to a point just above the bottom on the outer reef slope. When the fisherman feels sufficient activity on the line as a result of fish nibbling the bait, he pulls the *fen* up rapidly, usually catching several fish at once if he is skilled. The success of this method depends on the fact that triggerfish, in contrast to flying fish, swim downward when attempting to escape the net.

Since the invention of a special hook for triggerfishes (see Chapter 10) the *fen* has fallen out of favor on Tobi, for it requires considerable skill, as well as stamina, to use effectively. Traditionally a youth would not graduate to *fen* fishing until he was in his mid twenties and would not fully master the method until well into middle age.

Imported Fishing Techniques

Various fishing techniques used in the South West Islands are the result of contact with other cultures. Around 1890, for example, the famous Boston trader Captain David (His Majesty) O'Keefe persuaded some South West Islanders to work on his copra plantation on Mapia with the promise that he would return them to their homes after several years.[10] The islanders learned from the people of Mapia how to fish for tuna at depths of as much as ninety fathoms using droplines and hooks baited with fish. Prior to this tuna had been caught mainly by trolling close to the surface.

Tobians catch squirrelfish at night by torchlight using small feathered lures. The technique was taught to them, according to Patris, by some Indonesians who said that they in turn had learned it from Polynesians. (Kapingamarangi and Nukuoro islands are situated in central Micronesia but are inhabited by Polynesians who may have been the source of this technique.) More recently, in about 1945, spearguns and diving goggles were also introduced from Indonesia, but spearfishing has not achieved the importance in the South West Islands that it has in Palau.

Long ago, it is said, drift voyagers from Indonesia also brought the method of harvesting reef fish using a poison from the nut of the *Barringtonia* tree, a technique used throughout much of Oceania (Gatty, 1953).

10. This arrangement worked well for many years until O'Keefe and his ship were lost at sea in 1901. In 1956 a Trust Territory ship brought home nine descendants of the Sonsorolese left on Mapia.

Holden (1836) reported that, many years before him, a man from Ternate in the Moluccas had arrived on Tobi in a drifting canoe and had introduced a variety of religious innovations before leaving on a British schooner. According to Patris he also introduced a new fishhook design that the Tobians named after him. Another fishhook design came at an earlier date from another island near Ternate. Both designs are still in use today (see Chapter 10).

Just as trout fishermen match their artificial flies to the insect hatch, South West Islanders matched their artificial lures to the stomach contents of the tuna they caught. Native birds yielded only black, grey, or white feathers. The red jungle fowl, *Gallus gallus*, from which our domestic chicken was derived, is found in Palau where it was probably introduced in the distant past by voyages from Southeast Asia. But it never reached the South West Islands. When some of its domestic descendants were left there by a passing English ship in the nineteenth century, the islanders were more impressed with the utility of the feathers than with the flesh beneath them. Reddish brown chicken feathers were, and still are, considered particularly valuable, for they tend to match the color of squid on which tuna often feed. In the South West Islands, therefore, these feathers represent considerably more food when attached to a tuna lure than to a chicken. Not realizing the importance of chickens in this unaccustomed context, and that, when it comes to fishing, one chicken goes a long way, a number of observers have concluded that these birds were of little value to Pacific islanders (e.g., Eilers, 1936; Kotzebue, 1821; Lessa, 1975).

Because fishing conditions in the South West Islands differ greatly from those in Palau it can be seen that knowledge afforded by studying fishing and marine lore in the two island groups also differs. The next two chapters concern additional features of South West Island fishing that yield knowledge of the sea and its inhabitants that is largely unknown in nearby Palau.

CHAPTER

ISLAND CURRENTS

We were sailing to the island of Sonsorol on the schooner *New World*. It was shortly before 2:00 A.M. on a quiet, moonless night. My watchmate was Mariano Carlos, a young Sonsorolese chief home from law school in the midwestern United States. The captain had told us to wake him at 2:00 A.M., having reckoned that we would be close to Sonsorol by then and would have to drop the sails and wait for daylight before making our approach.

Without any warning the boat began to shudder and shake. A few seconds later, and with equal suddenness, the shaking stopped.

"Sonsorol is just ahead," said Mariano matter-of-factly. "It's time to wake the captain."

"How can you tell?" I said. "And what just shook the boat?"

"A current that occurs close to the island," he said.

This startling encounter with rough water in the middle of a calm sea prompted me to ask fishermen on Sonsorol, Tobi, and Pulo Anna to describe the currents around their islands. The three sets of accounts they provided were almost identical, yet they involved current patterns different from any of which I had previously heard (figure 3).

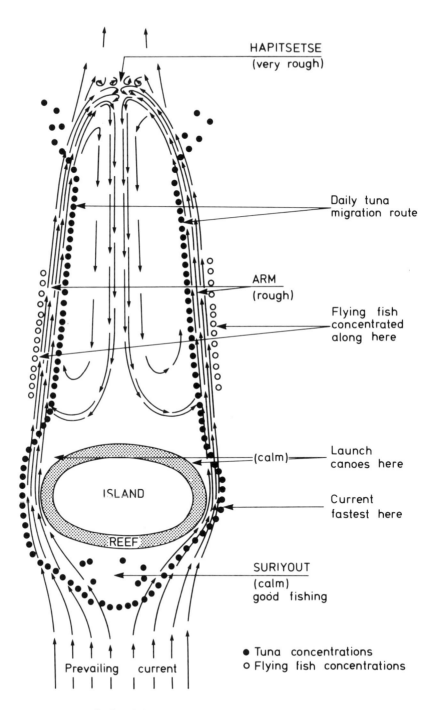

FIGURE 3. Idealized diagram of current patterns around Tobi.

On the upstream end of an island the current divides and is directed past the island on either side. Shoreward of the point where the current divides is an area of still, calm water with almost no current which Tobians call the *suriyout*.[1] Two narrow streams of turbulent water called *arm* extend downstream from the sides of the island, paralleling the prevailing current. Five to forty meters wide, each *arm* is sufficiently rough so that crossing it in a canoe can be hazardous. It was an *arm* that we had passed through near Sonsorol. These twin currents usually increase in width and decrease in turbulence with distance from the island. They extend downstream for a variable distance, sometimes to a point from which the island can no longer be seen, gradually curving toward each other. Ultimately they converge, creating an area of exceptionally rough water, called *hapitsetse*.[2] (If such a convergence can be found near Pulo Anna, fishermen there were unaware of it, although their description of the currents around their island coincided with those described for Tobi and Sonsorol in other details. Tobians said their *hapitsetse* disappears when the currents are weak.)

Leading back from the *hapitsetse* toward the island is a kind of backwash, or reverse current, running between the two *arm*. Patris explained it like this: "You know when you throw out a cigarette behind your canoe when you are sailing? Sometimes instead of being left behind it is sucked forward toward the stern. The same thing happens in the current directly behind Tobi. After we have been fishing downstream from the island we ride back to it easily on this current." Like fishermen, he said, turtles also tend to beach on the downstream side of the island, riding this backwash toward shore.

The wake current system as a whole is known on Tobi as the *hasetiho*. What did physical oceanographers have to say about such a current system, I wondered. Consulting the literature when I returned

1. Although parts of this chapter are based on information from Sonsorol, Tobian terminology is used throughout for the sake of clarity. Although inhabitants of the South West Islands all speak the same language, each island has a distinctive dialect. For example, *suriyout* is referred to as *doriyout* in the Sonsorolese dialect.

2. According to Mariano, a Japanese fisherman living on Sonsorol before World War II once approached the *hapitsetse* too closely in his canoe and was swept into a whirlpool. While clinging to the bow of his upended canoe as it revolved, he was spotted by a Sonsorolese fisherman who paddled his canoe as near as he dared. As the other man swung by he leaped from his canoe and grabbed the bow of his rescuer's boat. The Sonsorolese man back-paddled as hard as he could and the two men made it to safety.

Marine biologist John Miller (1973) described dye studies downstream of the islet of Molokini in Hawaii, the results of which indicated the existence of a pair of currents which converge in a manner similar to *arm*. Because of the exceptional turbulence at this spot, it is known to local boaters as The Washing Machine (Gene Helfman, pers. comm.).

to the United States I learned that they had reported two basic types of wakes downstream of oceanic islands (figure 4). One consists of a trail of more or less random turbulence. The other, called a von Karman vortex street, consists of a trail of eddies. The eddies form near both lateral edges of an island, enlarge as they drift downstream, and are shed laterally, first from one side of the wake, then from the other in a regular sequence. Neither of these wake types corresponded to what South West Island fishermen had described to me.

If no such current pattern existed, why had informants from islands separated by more than 100 miles of open sea been so uniform and explicit in their accounts of it, even possessing special words in their dialects for its features? Later, in a recently published thesis in anthropology, I found additional evidence for its existence. Fishermen from Linungan Island in the southern Philippines had

FIGURE 4.   Current patterns caused by an obstruction such as an island in a prevailing current. (a) von Karman vortex trail; (b) Random turbulence; (c) Stable eddy pair.

a.

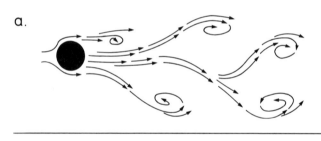

b.

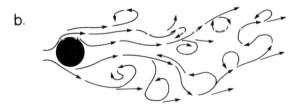

c.

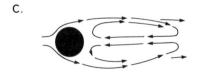

described the same current pattern around their island. Like South West Islanders they had specific terms in their language for: (1) paired currents that form on either side of the island; (2) a region in which these currents converge downstream; and (3) a back current flowing toward the island from this convergence point (Randall, 1977).

Perplexed, I decided to look into the literature on hydrodynamics to see if laboratory studies threw any light on the problem. Here I found that considerable research had been done on the influence of obstructions on water currents in experimental streams and *three* basic types of wakes had been observed. Two of them corresponded with those described by physical oceanographers. The third wake type, known as a stable vortex pair, corresponded with the *hasetiho* (figure 5).

Hydrodynamic theory predicts that the form a wake will take is a function of the size of the obstruction and the velocity of the current. The South West Islands present obstructions to prevailing currents ranging from about one to three miles in effective width at the surface. Prevailing current speeds in the area range typically up to two knots (Sailing Directions for the Pacific Islands, 1964). Current speeds between one and two knots plus island widths of one to three miles should produce stable eddy pairs according to theory (Richard Barkeley, pers. comm.).[3]

The reports of South West Island fishermen could thus be reconciled with scientific observations after all. The islanders had discovered stable vortex pairs and used them in their fishing and navigation long before they were known to science.

An account of how these currents relate to fish behavior and fishing activities in the South West Islands requires a brief description of seasonal changes in weather and sea conditions. On Tobi the year is divided into two roughly equal seasons based on current strength and wave conditions. The "peaceful" season, *Neihahi*, extends from February to September. (For part of this time Tobi is in the doldrums and the winds are very light.) The rough season, *Niyafang*, occupies the rest of the year. During this time canoe fishing is more difficult.

Sonsorol is far enough north of the equator to be in the monsoon belt, experiencing easterly winds from November to April and westerly winds from May to October. Sonsorol also experiences

---

3. Hydrodynamic theory does not predict the existence of quite such narrow, discretely defined streams of turbulence as the *arm*. But I experienced one in the dark, as recounted above, and Peter Black tells me that he often saw them while out with Tobian fishermen. Differences of scale between hydrodynamic laboratory models and currents around ocean islands may account for differences in the details of their stable wake patterns.

stronger currents and higher waves than Tobi and fishing conditions there are more difficult and dangerous.

The *hasetiho* shifts around these islands seasonally in synchrony with shifts in the direction of the prevailing currents. These seasonal shifts are predictable, and the fishermen know them well because their fishing sites and canoe-launching sites must shift with them. Around Tobi the current direction usually shifts clockwise with the seasons. The seasonal shifting of the current around Sonsorol is more erratic. Around both islands, at some times of year, the prevailing current may also fluctuate significantly in direction and speed within the space of as little as an hour. Such short-term fluctuations are more common and of greater magnitude around Sonsorol.

Two factors determine local current strength according to fishermen. The most obvious of these is the speed of the prevailing current. As the local currents flowing past either side of the island increase in speed, the *arm* get wider, rougher, and longer and the *hapitsetse* gets larger and rougher and moves farther downstream. But the direction of the current is also important. The wider the obstacle, the greater the volume of water that is deflected and the stronger the current this creates. Thus when prevailing currents hit the broad side of an island they accelerate more as they pass it than when currents of the same strength pass the island parallel to its long axis.

Tobian accounts of seasonal changes in local currents compare well with published accounts of major ocean currents in the region (e.g., Hisard et al., 1969; Schott, 1939). During the northern spring the westward-flowing South Equatorial Current is deflected north of the equator by New Guinea and flows past Tobi. Later in the year this current retreats southward. During May, June, and July, Tobi is often located near the boundary between the South Equatorial Current and the Equatorial Counter Current, and currents around the island are weak and exhibit considerable day-to-day, as well as year-to-year, fluctuations in direction. (In about two out of three years the currents set predominantly to the southwest at this time, in the other years mainly to the northeast.) By fall Tobi comes under the influence of the eastward-setting Equatorial Counter Current.

Currents around Tobi are not synchronous with those of Sonsorol, 150 miles to the northeast. In the summer, for example, Sonsorol is near the center of the Equatorial Counter Current and the currents are stronger then they ever get at Tobi, according to fishermen who have lived on both islands. Drawing a picture of the central Pacific Ocean on the sand with a stick, I once explained the large-scale seasonal shifts in equatorial currents to some South West

TABLE 5. Seasonal Changes in Weather, Wave, Current, and Fishing Conditions around Tobi.

Month	January	February	March	April	May	June	July	August	September	October	November	December
Approx. Tobian Equivalent	Tahebor	Yahemaus	Tumuch	Masichik	Masirap	Tauta	Tukumar (Tirotamau)	Huh	Ur	Erir (Yoruyoru)	Mar	Iich
WINDS AND WEATHER	Northeasterly, weak		Easterly or no wind, thunder	Easterly, weak, clear			Easterly, or no wind, rain, thunder		Northwesterly, moderate, dry			
CURRENTS	Southeast, weakening	South, weak	Southwest, very weak	Currents weak and unstable, direction often changing with the tide			South to east strengthening		South to east, strong			
WAVE STATE	Diminishing			Low				Picking up		High		
SKIPJACK & YELLOWFIN TUNA	Mostly large yellowfin, feeding on squid and flying fish		Small yellowfin mixed with skipjack move offshore before dawn, spread out all around Tobi in shallow water, returning in early evening eating small fish				Remain in Suriyout during day eating small fish, move offshore at night	Best tuna fishing season	Mostly large yellowfin, feeding on squid and flying fish			
JUVENILE REEF FISH		Particularly abundant				Abundant						
FLYING FISH AND NEEDLEFISH								Abundant, eggs appear in September. Spawning from October through June				
WAHOO			Very abundant, with eggs					Abundant				
GREEN TURTLES (Helen Reef)	Mating starts			Egg-laying. Numbers of eggs laid per nest decreases as the nesting season progresses, from 100-150 in April to around 70 in October							No turtles	
SEABIRDS			Nesting: Brown noddy, black noddy, white tern						Lesser frigate birds common; no nesting (nests on Sonsorol)			
DRIFTING LOGS							Some logs		Best log months			

Island fishermen. They seemed as interested in this glimpse of the larger pattern as I was in the details of the local circulation they had taught me.

Fishermen of all three islands noted that in some years the prevailing currents behave atypically. For example, in Tobi the direction of the current occasionally shifts systematically counterclockwise for several months. Oceanographic observations in this region confirm that current patterns differ appreciably from the norm in some years (e.g., Inanami, 1941).

Prevailing currents around Tobi are weakest from April through February. Pelagic juvenile reef fish are also most abundant around the island during this time. (The apparent connection between these two phenomena is discussed in Chapter 3.) Tobians are particularly aware of seasonal changes in abundance of juvenile reef fish because of their importance as food for tuna. Few schools of small, surface-feeding tuna, and few seabirds, are present from September through January. The return to Tobi waters of large numbers of tuna and seabirds beginning in late February is believed by fishermen to be linked to the increased availability of pelagic juvenile reef fish. Egg-bearing *yar* or wahoo (*Acanthocybium solandri*) are also more abundant around all three South West Islands (and Palau) in the spring.

Mackerel tuna (kawa kawa) and dogtooth tuna remain near the islands at all times according to fishermen. But yellowfin tuna, skipjack tuna, and a large needlefish all make predictable daily migrations to and from the waters near the islands, and these migrations vary in character with the season. During January and early February at Tobi, for example, skipjack, yellowfin, and needlefish move offshore after midnight and return by mid-to late afternoon. Skipjack move offshore and return earlier than yellowfin. Needlefish tend to leave and return earlier than either of the tunas. The phase of the moon reportedly has some influence on the timing of these movements; fish are said to leave the island later during the dark nights around the new moon and return later.

These daily migrations take tuna and needlefish from the calm waters of the *suriyout* along the *arm* in the direction of the *hapitsetse*. Seabirds leaving the island at dawn to hunt for the small fish that the tuna drive to the surface also generally fly off in this direction. Both birds and fish also generally return from this direction.

From February through June the tuna move offshore before dawn and are found during the day spread out near the surface all around the island. During this period of weak prevailing currents around Tobi, the schools of small fishes on which the tuna feed are also spread out around the island. When the prevailing current picks up speed in July these forage fish tend to stay closer to the island and to the *arm* and form more compact schools. This, say the fishermen, causes the tuna to concentrate in these areas too. In July and August the tuna tend to concentrate in the *suriyout*[4] and stay there throughout the day, moving offshore at night. In the fall large yellowfin dominate the catch. They tend to stay in deep water, moving offshore about 3:00 A.M. and returning in the mid-afternoon. Large, deep-running trolling lures are designed for use at this time. Yellowfin are also caught by dropline fishing to depths of 100 meters or more in the reduced currents of the *suriyout* at this time.

Small species of needlefish and flying fish can be found in the *suriyout* throughout the year. Larger and more sought-after species of both types are found further offshore. Large flying fish are particularly abundant near the outer margin of the *arm*. Small yellowfin and skipjack tend to concentrate near the inner margins of the *arm* as they move to and from the island.

Similar patterns of movements of offshore–inshore movements were described to me by fishermen from Sonsorol, Pulo Anna, and outer islands in the Ponape district in the eastern Carolines and in Palau by Ngiraklang. The seasonal timing of these movements varies from island to island, presumably because of seasonal differences in current and wind conditions.

I searched the literature to find the extent to which scientific research might support the reports of the fishermen on daily tuna migrations. Much work has been published on long-distance seasonal migrations of tuna in the Pacific. But I could find only one paper that relates to daily, repetitive migrations near islands. Marine biologist Heeny Yuen (1970) fixed an ultrasonic tracking device to a skipjack tuna near Hawaii and monitored its movements. When returned to the water the fish rejoined its school. Each night the fish left the shallow bank where it was caught and traveled over deep water for distances of up to sixty-six miles before returning to the bank by dawn. The pattern of movement described by Yuen on the basis of observations on a single fish is the same as that described by fishermen for populations of fish of three different species. Their testimony indicates that these movements vary seasonally, are related to prevailing current strength and direction, and are not just a local Hawaii phenomenon.

4. Westenberg (1953) similarly notes that in Indonesia favored tuna-fishing spots are where "a prominent corner divides the approaching water masses of the surface current."

C H A P T E R

FISHHOOKS

9

No matter how good . . . the rest of your tackle, it is that little hook that holds and lands your fish. Or does not hook him because it is badly designed.

—*R. L. Bergman*

Traditionally Micronesians fashioned their fishhooks from bone, shell, or wood. Forming and finishing such hooks using nonmetal tools was a laborious task often taking several days. With so much labor going into their making, and so much hinging on their successful functioning, such hooks were valued much more highly than we value modern hooks. Accordingly they were fashioned with considerable care. In Palau the knowledge that went into the design of such hooks has vanished; even Ngiraklang knew nothing about them. His hooks came from the store and were made in factories thousands of miles away. Such was also the case with most of the other fishermen I met throughout Micronesia.

But in the South West Islands the older fishermen still prefer to make their own hooks. They are now made of stainless steel wire (some of it coming from Japanese war planes downed during World War II). But they are formed by hand, using grooved rocks as anvils, and chunks of cast-off metal as hammers. And most of the knowledge that goes into their shaping predates the invention of stainless steel by centuries.

The preservation among Tobians of the ancient skills involved in the design and use of such hooks provided a rare opportunity to examine their function. Such a study is of more than academic interest, for the question of the relationship of shape to function in fishhooks has received little attention from marine researchers (e.g., Baranov, 1976; Saetersdal, 1963); and the form of many machine-made hooks owes as much to fashion as to proven function (Hurum, 1977).

Hundreds of pages have been written about the fishhooks of Oceania by anthropologists, and it might be thought that much could be learned from them about functional design. But hooks have been used in these studies mainly as aids in establishing cultural sequences and historical links, much like pottery. Comparatively little effort has been made to understand what their makers undoubtedly considered to be their most important attribute—their ability to catch fish.[1] So little is known in fact that Reinman (1967, p. 188), in a major review of fishing archaeology in Oceania, states with elaborate caution, "although we are here entering *an area of speculation*, there is *some* evidence *suggesting* that *perhaps* the different shapes have *some* significance in terms of the kind and size of fish taken and do not *just* represent *idiosyncratic behavior* on the part of the makers" (emphasis added).

Some early European observers scorned these stone, bone, sea-shell, wood, or coconut shell devices. "Ill-made," said Banks (cited in Best, 1929). "Very clumsy affairs," said Polack (cited in Best, 1929). "Ill-contrived for the purpose," said Horace Holden (1836, p. 38) of Tobian turtle shell fishhooks.

What then should we make of the fact that other Europeans, after having tried native hooks, favored them over European designs? One admirer described them as "a triumph of stone age technology" (Beasley, 1928, p. 20). Captain James Cook (1785, pp. 149, 150) said of Hawaiian hooks, "considering the materials of which these hooks are made, their strength and neatness are really astonishing; and in fact we found them, upon trial, much superior to our own." U-shaped European hooks are "far inferior" to native hooks, stated Danielsson (1956, p. 183). "My experience is that the native form of hook is preferable," claimed an early New Zealand visitor. "I always made my own *hapuku* hooks" (Best, 1929, p. 36). "After a short sojourn in the Pacific I gave up European and took to native fishhooks, and always found the latter more deadly," said Romilly

---

1. Nordhoff (1930) fished extensively with the Tahitian pearl shell tuna lure, however, and carefully described the functional aspects of its design. Kennedy (1929) provided a useful description of the function of a specialized Pacific island hook used to catch the oilfish, *Ruvettus pretiosus*.

FIGURE 5.   How a rotating hook functions. See text for explanation.

(1856, p. 133). Of Tahitian hooks Wilson (1759, p. 386) said, "notwithstanding the form which to us appears most clumsy and crude, they will succeed, when we with our best hooks cannot." "In my opinion the Polynesian incurved hook . . . is mechanically superior to hooks of our kind," said Nordhoff (1928, p. 44). "The old Hawaiian fishermen caught more with these peculiar hooks than they could with the more dangerous looking hooks of the foreigner," stated Brigham (1903, p. 8).

Adding to this testimony are numerous reports that Pacific islanders had little use for metal fishhooks of European design, reshaping them and often removing the barb. Ellis (1859, p. 150), for example, claimed that the Tahitian fisherman "would rather have a wrought iron nail three or four inches long, or a piece of iron wire of the size, and make a hook according to his own mind than have the best European-made hook that could be given to him." Even a relatively fragile pearl shell hook was considered "much better than any made in Europe," Ellis stated.

An examination of Tobian fishhooks helps explain these conflicting accounts. Of the thirteen basic hook patterns in use on Tobi, the invention of all but one appears to predate the introduction of metal to Oceania. And when the older Tobian fisherman does come into the possession of an imported hook, he sometimes modifies it strongly before use, just as his ancestors did in the nineteenth century.[2]

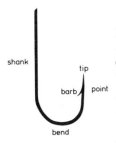

shank

tip

barb   point

bend

The feature of many Pacific island hooks that most often perplexes the Westerner is the strongly in-curved point. The tip, instead of pointing upward, points inward toward the shank, or even downward (figure 5 and plate 21). Western fishermen unfamiliar with such hooks instinctively doubt their effectiveness. When noted American angler Harlan Major first saw one, for example, he was skeptical that "it would catch—to say nothing of hold—any fish" (Major, 1939, p. 52). But the design is intentional and the hook highly effective when employed properly. Called a rotating hook,[3] it is used under conditions where it is difficult to set the hook by jerking the line sharply when a fish bites. Such conditions occur when dropline fishing either in deep water or in strong currents. In either case water currents cause the line to "belly," or curve, rather than hanging straight down. Jerking the line tends simply to straighten it somewhat and little of the energy applied is transferred to the hook.

2. According to Holden (1836), Tobian fishermen could not be induced to use the European hooks his crew gave them "till they had heated them and altered their form."

3. Nordhoff (1930) learned how these hooks functioned and appears to have been the first writer to use the term "rotating" to describe them.

The fisherman allows a fish that has taken the bait under these conditions to swim with it for a time. As the fish swims away from the fisherman and roughly parallel to the line, the hook is usually pulled to the corner of the jaw. When this happens the point catches on the flesh of the inner edge of the upper jawbone and serves as a pivot point for the hook. Increasing tension on the line meanwhile causes the shank to compress a portion of the upper jaw (figure 5a). As line tension causes the hook to pivot, the jaw is squeezed between point and shaft and forced through the gape of the hook into the bend (figure 5b). This releases the pressure of the shank on it and causes the compressed portion of the jaw to expand to its normal dimensions. The incurved point now holds the fish on the hook; the diameter of the flesh and bone contained within the bend is now greater than the width of the gape of the hook. The jaw would have to be compressed once again in order to slip back out through the gape. There is no simple means by which the struggling fish can bring to bear the appropriate pressure to accomplish this. As the fish turns more or less at right angles to the line to fight the increasing pressure, the hook rotates in the jaw, forcing the point completely through it (figure 5c). The line is generally attached to the inner side of the shank to favor this rotating motion.[4]

Because turtle shell hooks were somewhat flexible the gape would widen slightly when a fish pulled against the line, making escape somewhat easier than is now the case with metal hooks. Consequently Tobian rotating hooks made of turtle shell often had barbs to offset this tendency. Tobian stainless steel rotating hooks have no barbs. Tobians, like other Pacific islanders, often take the barbs off manufactured metal hooks when reshaping them into rotating hooks. There are three reasons for this. First, the barb reduces the ease of penetration of the point. Second, as the fish fights, the barb tends to tear at and widen the hole through which the hook penetrates the jaw, thereby increasing the likelihood of the hook pulling out. Third, the barb constitutes an impediment when a fisherman unhooks his fish. Rotating hooks hold fish very well without a barb. In fact one disadvantage of a rotating hook is that it is so difficult to unhook, even when unbarbed, that it often has to be cut out of the fish (see also Kawaguchi, 1974).

The greater the slack in the line the less control the fisherman has over the fish, and the greater the need for a hook that will hold a fish securely in the absence of tension. Most rotating hooks on Tobi are therefore made in two different models depending on the depth at

---

4. Nordhoff (1930) made the same observation with regard to Tahitian line attachments.

which they are to be used. Although both models are basically deep water hooks, the model with the point closest to the shaft is used in the deepest water when the belly in the line will be greatest and the line slackest. A variable compromise is thus made between the speed and ease with which a fish can be hooked (faster and easier when the gape is wide) and the likelihood of retaining the fish on the hook (greater when the gape is narrow). Deep water fish are worth the extra time and care involved in hooking them. The deeper Tobians fish, the larger, on the average, are the fish they catch.[5] They fish to depths of more than 600 feet.

Also influencing the fisherman's choice of the particular type of rotating hooks to use is the number of sharks in the vicinity. When sharks prove troublesome, a hook with a wider gape is used. Fish escape more easily from such a hook but they hook up faster. This reduces the time they must be played and during which they make easy targets for sharks.

The range of compromise between the speed and the secureness with which a fish can be hooked is further extended by another basic fishhook type, the "jabbing" hook. This is a shallow water hook used with a taut line in trolling, shallow dropline fishing, and pole fishing and is of the general design most familiar to Westerners. There is little if any inward curve to the point and the gape is generally greater than that of a rotating hook. A fisherman jerks a taut line to set this hook. Only the barb, if it is present, plus sustained tension on the line, prevents a fish from throwing a jabbing hook. And as the hole in the jaw created by such a hook becomes enlarged during the struggle, the barb holds the fish progressively less securely. The fisherman must thus land the fish as quickly as possible.

Whether a jabbing hook is used with or without a barb often depends on how fast the fish are biting. Tuna in a school feeding at the surface often bite voraciously. But the school usually dives after a few minutes, so as little time as possible must be wasted unhooking fish. In this kind of fishing the fisherman's skill is measured largely by the speed with which he lands and unhooks fish and gets his hook back into the water. Concerning similar fishing in Tahiti, Nordhoff (1930, p. 215) states:

> The skill of a bonito-fisherman may be judged from an inspection of his hooks. A green hand uses long points, very sharp, to ensure landing every fish that strikes. The expert uses short, blunt points, just sharp enough to lift the fish out of the water before they drop out of the jaw.

---

5. An increase in mean size of fish with depth has often been noted by biologists (Helfman, 1968), but no one as yet has come up with a generally accepted explanation for this trend.

While the beginner is landing a dozen bonito, many of which must be disengaged from the hook by hand, the adept will have pulled out of the water fifty fish and landed forty-five of them without touching the hook.

Under such conditions many Pacific island tuna fishermen, including Tobians, traditionally used barbless hooks, as do tuna pole fishermen of countries throughout the world today.

A barbless hook has the added advantage of penetrating the fish's mouth faster and with less resistance. Barbless jabbing hooks are also used in the South West Islands for pole fishing on the reef; a fish has little time to struggle and escape before being yanked from the water when hooked under these conditions. The same hook type is used for shallow dropline fishing when the fish are biting rapidly and, as in the case of fishing for surface-feeding tuna, speed of unhooking is important. Much line fishing is done at night on Tobi. The ease with which an unbarbed jabbing hook can be removed from a fish is also an advantage on a moonless night when the fisherman cannot see clearly what he is doing.

There are nine basic types of Tobian jabbing hooks. All are for use in shallow water. As with rotating hooks several of these types come in two different models. Once again the model with the point closer to the shaft is used with a slacker line than the other model with a wider gape. But whereas the point is curved inward to bring it closer to the shaft in a rotating hook, the shank is bent to bring it closer to the point in a jabbing hook. The straight-shanked version is used for trolling close to the canoe—a situation in which the line is almost always straight. Bent-shanked jabbing hooks are used in trolling with a long line and for shallow dropline fishing—situations in which the line is more curved and the tension more variable.

Until recently only the subjective impressions of Pacific island fishermen and their European converts could be used to support the contention that incurved hooks are superior to those with straight points. But recent trials conducted in the Caribbean (Kawaguchi, 1974), in Scandinavia (Hamre cited in Hurum, 1977), and in Great Britain (Forster, 1973) have demonstrated that in dropline fishing incurved hooks catch more fish.

The incurved points of rotating hooks also render them less liable to hook up on the bottom—an important feature when dropline fishing over coral-studded bottoms. Before the arrival of metal, rotating hooks had an additional advantage over jabbing hooks. Bone or shell hooks could not be jerked as hard as metal hooks for fear of breaking them. Therefore a hook that imbedded itself in a fish's mouth when only a gentle pull was exerted on the line was valuable even in shallow water, particularly when large fish were being sought.

Like many other Pacific islanders Tobians do not just "go fishing"; they fish for specific species and their techniques vary accordingly. There is a saying on Tobi that "the hands of a fish are its mouth"—meaning that in the absence of hands, fish manipulate their food with their mouths. Some hold their food between their teeth or lips and move some distance with it, often repositioning it before swallowing it. Groupers tend to "inhale" food deep into their mouths with a single motion. Others nip off small pieces. Still others take food into their mouths and blow it out repeatedly before swallowing it. Mouth shapes, sizes, and hardness also vary greatly in different fishes. All these factors are considered when deciding what hook to use for a particular species.

The wider the bend in a hook the greater the difficulty a fish has in spitting it out. Thus in the South West Islands hooks with wide bends are used for large-mouthed fish such as groupers and most squirrelfish. Such hooks tend to hook deeply in the jaw rather than the lips and are therefore also used when fishing for species with easily torn lips, such as rainbow runners and tuna. Smaller sizes of wide-bend hooks are used for smaller-mouthed fish, such as wrasse and fusiliers, that tend repeatedly to suck in and spit out the bait cautiously. Hooks with narrower bends are used for small-mouthed fish such as squirrelfish of the genus *Holocentrus*. Hooks with long points tend to be used for species with narrow, deep mouths, whereas short points are used for those with shallow mouth cavities. Another factor in determining point length is that a short point penetrates quickly but is likely to be dislodged by a species that struggles vigorously; a long point does not drive home as readily but is more difficult for a fish to throw.

Tobians distinguish a third type of hook—the *fong* hook. When the terminal few millimeters of a hook point are bent sharply inward this tip is known as a *fong*. Such hooks may function like rotating or jabbing hooks depending on their shape. The *fong* serves to help keep the fish on the hook. And in two out of three *fong* hooks it also helps keep the bait on the hook. The third *fong* hook, of unusual and ingenious design, was invented on Tobi after the introduction of metal. Called *haufong*, it was designed specifically to catch triggerfish, a group of fish which Tobians relish.

Triggerfishes have small mouths with exceptionally strong jaws (Aleev, 1969). Because they take small bites, many species do not take bait of normal size into their mouths whole. Instead, they nip pieces off the bait until what remains falls off the hook. They are, in consequence, notorious bait stealers (e.g., Buck, 1949). If, to counteract this, a very small hook is used with a small piece of bait, they will take the entire hook into their mouths, often snipping the line to which it is attached with their teeth.

fong

bait

Haufong

The *haufong* hook was invented on Tobi in the nineteenth century by a half-Tobian, half-Sonsorolese fisherman to circumvent these problems. A small portion of bait is placed on the hook so that it covers only the *fong*. (The tips of all other baited hooks are left showing.) Because the bait is small the triggerfish takes it between its lips in one piece along with the *fong* within it. When the fisherman pulls on the line the fish tries to spit the hook tip out, but the *fong* catches on its upper lip and is then driven through it. The fish then slides down the hook to the bend. The *fong* now functions as a barb.

The *haufong* is also the hook of choice for some other strong-jawed, small-mouthed fishes such as unicorn fish.[6] It is sometimes also used in fishing for other species such as fusiliers, which cautiously and gently suck the bait in and spit it out repeatedly; the *fong* tends to catch easily in their upper lip or jaw when they try to spit out the bait. Although a variety of wrasse have small mouths, strong jaws, and tend to mouth bait cautiously, this hook does not generally work well for them. Their lips are too small and their palatine bones tend to prevent the *fong* from penetrating the roof of the mouth easily.

Two different forms of *haufong* are shown on page 197. The "new" style was probably adopted because the point is directed more or less parallel to the line of tension created when fishermen pull against fish; the closer the line of penetration coincides with the line of tension the greater the penetrating power of the hook.

Patris explained to me that the *haufong* design would be of little use with hooks made of bone or shell. The *fong* would be too weak to withstand the strain put on it by a triggerfish. However a hook of similar design and function was invented independently in the Tuamotus 3,000 miles to the southwest of Tobi prior to the use of metal. Seurat (1905) states that the bait was placed on the extremity (i.e., the "*fong*") of this hook and that a version made of pearl shell was used to catch parrotfish. Because the strong jaws of triggerfish

6. Traditional bone, shell or wooden hooks of somewhat similar design were made in various parts of Oceania, but the bent tip was longer and thicker and served a different purpose. Large pieces of bait were placed on the hook *below* the bent portion of the tip, which served to prevent the bait from falling off, as well as decreasing the gape of the hook. Such hooks were used, unlike the *haufong*, for certain *large*-mouthed species. One such hook, the large *Ruvettus* hook (e.g., Kennedy, 1929) was not used in the South West Islands. The fish for which it was intended, *Ruvettus pretiosus*, apparently does not occur here; Tobians did not recognize photographs I showed them of this distinctive fish. As they routinely fish at a depth of several hundred feet on the outer reef slope (where this fish is found around many other Pacific islands), they would surely have caught it if it occurred around Tobi.

would break such shell hooks, wooden hooks of the same design were used to catch them. The shrub *Pemphis acidula*, which has very hard wood (Stone, 1970), was used for this purpose in the Tuamotus. This plant is very common and widespread throughout the coastal tropical Pacific and it is probable that it grows on Tobi. But unlike many Pacific islanders, Tobians do not appear to have used wood for making fishhooks.

Another unusual Tobian hook is called *man tanante. Man* means "being" or "person." *Tanante* is a corruption of the name Ternate, an island about 300 miles southwest of Tobi in the Moluccas. The hook name commemorates a man from Ternate who accidentally drifted in his canoe to Tobi many generations ago and introduced the design.

The design is unusual among metal hooks in that the shaft is scarcely longer than the point. I was not able to understand the functional aspects of its design, but was told that it is the Tobians' most versatile hook, and is therefore used in instances where the fisherman is unsure as to what kind of fish he is liable to catch. Once he determines which fish are biting he often switches to a hook more specific to that fish.

Traditional Pacific island fishhooks clearly suffer in comparison with Western hooks in some regards. Metal hooks have greater tensile strength. Metal and shell hooks of the same style are therefore of somewhat different shapes and proportions. Shell hooks are usually thicker throughout. Jabbing shell hooks are also often reinforced at their weakest point—the bend—with a kind of triangular keel, particularly if they were to be used on a lure for trolling for large fish. Shell hooks also tend to have shorter shanks than comparable metal hooks because of the greater likelihood of a fish biting through or snapping a long shank.

Because the manufacture of shell hooks was so time-consuming, they were treated with great care. On Tobi if a grouper ran into a hole in the reef with a hook, the line was not broken off and the hook sacrificed as it usually is today. Instead steady tension was kept on the line until the grouper finally emerged—sometimes as much as an hour later. If a hook got snagged on a coral, a rock was attached to a second line, hooked on the fishing line, and slid down it. A little slack was let out in the fishing line so that the rock weight would pull on the hook from below, thereby sometimes unsnagging it in situations where an upward pull was of no avail.

Despite the labor involved in making turtle shell hooks and the fact that Tobians have had enough metal to make metal fishhooks since before the turn of the century, they did not abandon the use of turtle shell hooks completely until the late 1930s. One advantage of the latter, according to Patris, is that many fish seem not to like the taste of metal. Once they throw a metal hook they are not soon likely to bite again. Bone or shell hooks, however, apparently taste "natural"; a fish is more likely to bite again on such hooks if not hooked the first time.

A device that operated much like a primitive bow drill (plates 22, 23) was used to cut hooks from turtle shell. (Unlike some Pacific islanders Tobians never used heat to mold their turtle shell hooks.) Instructed by the older men, Patris made one of these instruments for me—probably the first to be made in the South West Islands in many decades. The size of the hooks being cut could be varied by using an adjustable wooden wedge lashed between the cutting teeth and the pivot tooth. The drill was rotated gently and repeatedly to cut through the shell. The hook was then finished using a file made from coral and "sandpaper" made from the abrasive skin of the nurse shark.

Some turtle shell trolling hooks not only had a barb in the conventional position near the tip, but also a barb facing it, projecting from the shank. These hooks were used when one man trolled with two lines. Although the second barb made it harder to hook a fish, it made it easier to hold a hooked fish on an unattended line while the fisherman played a fish on his other line. Occasionally a second barb was placed facing outward near the tip for the same purpose. Paired barbs were never used on rotating hooks.

Superimposed on the thirteen basic Tobian hook designs (appendix C) are many intergradations in style, some with special names. In addition there are other subtle variations on these designs, all of which are considered by their makers to be of functional rather than stylistic significance. I was often unable to perceive these variations when examining the hooks, nor did I clearly understand their significance when it was explained to me.

If I had had more time and opportunity to fish with the Tobians so as to test all these designs and their variations thoroughly, I have no doubt that this chapter would be much longer. Nevertheless this brief survey makes it clear that we can go beyond Reinman's statement that "perhaps" different shapes and sizes of Oceanic fishhooks had "some" significance beyond the whims of their makers. Fishhooks of the South West Islands have been skillfully designed to take into account the size of the fish being sought, their mouth size and shape, their biting characteristics, the toughness of their mouths, the depth at which they are being sought, the strength of the current, and the presence or absence of troublesome sharks. The ease with which different types snag up on the bottom is also taken into consideration. A varying compromise in design is made between the ease with which a hook can be set in a fish and the speed with which it can be shaken loose by the fish or removed once the fish is landed.

But, lest it be concluded from this account that the design and use of fishhooks has achieved the status of an exact science in the South West Islands, it should be added that there remains plenty of latitude for difference of opinion among the fishermen as to what hook to use for a particular purpose and how best to form it.

Those who maintained that Pacific island hooks were crude and ineffective were misled by the strange shape of the rotating hooks; it is not intuitively obvious that such hooks are very effective if used properly. O'Connell, a castaway on Ponape, provides an example of this misunderstanding. Rejecting the "rude tortoise shell hooks" of the natives, he made some "very tolerable" hooks from the ramrods of muskets preserved from the wreck of his ship. "But it was necessary," he related, "to keep the line taut, as there being no barb, the fish would otherwise escape" (O'Connell, 1836, p. 112). Had he copied the design of the rotating turtle shell hooks he saw, and sought instructions from Ponapeans in their use, he would not have been inconvenienced by the lack of a barb.

Those, in turn, who maintained that the islanders' hooks (by which they meant rotating hooks) were superior to Western jabbing hooks were oversimplifying. Rotating hooks were indeed better in design (although of lesser tensile strength) than typical Western jabbing hooks for those types of fishing involving slack lines. But the jabbing hook, native both to Europe and the Pacific islands, is a useful and versatile hook provided that it is used when fishing with a taut line in shallow water.

Contrary to the impression that is given by the early literature, it is unlikely that the islanders always modified the shapes of European hooks before using them. They probably did so only when the type of fish or style of fishing for which they were intended required it.

Lines, Leaders, and Lures

Today South West islanders still fish often with lines they make themselves out of sennet (coconut husk fiber) or the inner bark of *Hibiscus tilaceus*. One observer noted in 1898 that Tobian sennet "was twisted in various thicknesses so prettily and with such regularity that a European cordmaker would have gained credit by it" (Eilers, 1936). But good sennet rope is more than pretty. It is not only strong but also exceptionally resistant to decomposition. Heavy deep-water fishing lines of sennet are valued more highly by their Tobian owners than commercial line and often last for several generations.

Tobians recount that on recovering sennet and nylon fishing line from caches made during World War II they discovered that the nylon lines had deteriorated badly whereas the sennet line was unaltered. One of the latter coils is still in use today. Kayser (1936) notes similarly that coconut fiber left by accident for twenty-two years in a "slime pool" on Nauru was still in good shape when recovered.

Seidel (1905) described Tobian hibiscus fishing lines as "the best of their kind to be had without machines." Because the supply of fiber hibiscus is not as great as that of coconut fiber, hibiscus is used only for shorter, thinner lines. For deep trolling several meters below the surface, hibiscus is preferred because it is less buoyant than coconut fiber. Owing to its greater strength it is also used in preference to coconut fiber for making leaders. Leaders must be strong yet thin so as to be inconspicuous to an approaching fish.

The ends of the shanks of Tobian turtle shell fishhooks were unique in that they were forked and, according to Beasley (1928), provided "a more practical method of attaching the snood (leader) than is in use in any other locality." Today the shanks of stainless steel hooks usually terminate in an open loop. The leader is attached first with a slip knot, then with two or more half-hitches running down the shaft, then with a final half-hitch in the loop, over the slip knot. Because the loop is not closed the line can be quickly loosened and slipped off in order to change hooks but will not slip off when a fish is hooked.

Wire leader is sometimes used today when fish with sharp teeth, such as sharks or barracuda, are being sought. Patris pointed out to me that there is a disadvantage to this arrangement. The attachment of the rather stiff metal leader to the line provides a second pivot point (the first one being where the leader is attached to the hook), allowing the fish greater flexibility in its struggles to escape. I showed him a piece of flexible braided metal leader I had recently bought and he pronounced it an improvement over the stiff wire leader with which he was familiar.

As on many other Pacific islands, trolling lures were made from various types of shell and coral adorned with feathers (figure 6). Whereas modern research has shown clearly that colors are perceived by some fishes and that some exhibit color preferences when feeding (e.g., Ginetz and Larkin, 1973; Wagner and Wolf, 1974), the claims of some Pacific islanders that subtle differences in the coloration of tuna lures make big differences in their effectiveness is not wholly convincing. Much has been written, for example, about the importance of shading in pearl shell tuna lures (e.g., Buck, 1932; Nordhoff, 1930). But Tobians say that except when tuna are feeding on squid, when a reddish-brown lure is preferred, tuna show little color preference.

The experiences of Palauans and of marine biologists tend to support this testimony. Today when feathers are torn by fish from worn commercial tuna lures, Palauans customarily replace them with skirts cut from plastic shopping bags. The color of the skirts may be white, grey, pink, purple, green, or blue with varying amounts of black print. Palauans express little preference for one color over another and state that the fish seem similarly undiscriminating. Ommaney (1966) experimented in the Indian Ocean with trolling lures made of bone, shuttlecock, feathers, plastic strips, rope yarns, metal foil, metal spoons, blobs of lead, and tops of cigarette cartons. "But," he states, "we never really found that one sort of lure was decisively better than any other. When the fishing was good any kind of lure would do as well as any other." In tank experiments Hsiao and Tester (1955) showed that to kawa kawa (mackerel tuna), *Euthynnus affinis*, yellow, red, black, white, and combination lures were almost equally attractive, although there may have been a slight preference for white.

FIGURE 6.   Tobian trolling lure, for smaller, near-reef species such as jacks. The body is made of red gorgonian coral decorated with chicken feathers. Over the feathers one scale of a large parrot-fish is tied in order to minimize the damage done by the teeth of a struggling fish. The hook is made from turtle shell. The lashings and leader are made from *Hibiscus* fiber.

CHAPTER

SON OF THUNDER AND OTHER FISHES

10

Fish species are conceived of by Palauans and South West Islanders as belonging in groups or families. Species in the family we call *Scaridae*, or parrotfish, for example, are called *butiliang* in Palau. Each important species within such a group has a name that distinguishes it from others in the group. Thus bumphead parrotfish, *Bolbometopon muricatus*, are called *kemedukl*. Similar systems of binomial nomenclature characterize native naming systems for plants and animals throughout the world.

The resemblance between these naming systems and those used by biologists has often been noted. The similarities are not as great as they might appear, however. The two-name scientific system serves two main purposes. One is to designate the common evolutionary origin of members of groups of closely related species, such as trout or pine trees or rats. The second is to provide a naming system that is uniform throughout the international scientific community. This is essential in order to avoid the confusion that would result if biologists from different areas tried to communicate with one another using the local popular names of species.

Sometimes there are no popular names, as in the case of many rarely seen deep-sea fishes. Other groups, such as the cardinal fishes, have only one common English popular name although there are dozens of different species. In addition, popular names vary not only from language to language, but also from region to region with areas where a single language is spoken. For example, the fish known as cutthroat trout where I grew up is known in other parts of western North America as Great Basin trout, Montana black-spotted trout, Columbia River trout, Colorado River trout, and so on. But to biologists from Siberia to Pretoria, whatever their language, this species is identified unmistakably by the name *Salmo clarki*.

The functions of fish-naming systems in Palau and the South West Islands are somewhat different and their structures more flexible than those of the system used by biological taxonomists. Many groups of fish are given a common generic name because of their anatomical similarity, just as they are by biologists. But others that biologists recognize as belonging to a single group on the basis of their anatomy are divided by native fishermen into separate groups on the basis of other characteristics. On Tobi, for example, bright-colored groupers (serranids) are distinguished from other groupers by the generic term *hari*. An overlapping generic term for certain groupers is *bwerre*, which refers to the multihooked dropline method of fishing used to catch them. Such overlaps are not permissible in formal biological taxonomy, which has to encompass all the world's 20,000 species of fishes. Otherwise substantial confusion would result. But Tobian taxonomy need not be so rigorous because Tobians differentiate between only about 200 species of fish.

Snappers, when they are very small, and damselfish (which never attain a large size) are known collectively among Tobians by the generic name *richoh*. These two groups are not closely related in an evolutionary sense. But they are closely related in Tobian minds, and quite logically so. They look fairly similar, taste similar, live in similar habitats, eat similar food, and are caught by the same methods.

Biologists devote as much attention to the naming of rare species as they do to naming common ones. Pacific islanders, in contrast, give specific names only to those species that are important or otherwise interesting to them. Cardinal fishes, for example, are unimportant to Tobians. Consequently, although there are probably at least a dozen different species of cardinal fishes around Tobi, they are all known only by the name *hauborap*.

In contrast, a species that is very important to Pacific islanders may have a variety of names, often corresponding to different size classes. In most cases it is recognized that these different names refer to the same fish. But the different sizes are distinguished taxonomically because of their different habits or habitats or the different fishing techniques used to catch them (e.g., Gosline and Brock, 1960; Helfman and Randall, 1973; Ottino and Plessis, 1972; Randall, 1973). In Palau, for example, the milkfish, *Chanos chanos*, is called *chaol* when it is small and lives in brackish mangrove ponds. After growing larger it moves out onto the reef to live over sandy bottoms and it is now called *mesekelat*. (Most Palauan fishermen are aware that the two names refer to different stages in the life history of the same species.) Once snappers become big and move into deeper water, Tobians no longer call them *richoh*. The dull-colored species are now known by the generic name, *hatih*, which means "hard to find" and refers to the fact that their dull coloration makes them hard to see in their natural habitats. Both dogtooth tuna and dolphinfish have three different names on Tobi relating to different size classes.

I asked the Tobians to help me make a dictionary of Tobian fish names in order to facilitate our communication during interviews. Later it occurred to me that something might be learned about various fish by asking for the derivations of these names. I found that some names had no remembered meaning apart from that of designating the fish. Such names correspond to English names such as "eel" or "mullet." But many Tobian fish names, I was told, consist of syllables or phrases describing some characteristic of the fish they represent, thus corresponding to English fish names such as "flying fish" or "grunt."[1]

Some Tobian names refer to the biting habits of the species in question. *Hari*, the generic name mentioned above for bright-colored groupers, means "always bites." One of these groupers, *Cephalopholis cyanostigma*, is apparently particularly hook-prone, for its full name, *Hari merong*, means, loosely, "always bites, takes any bait." *Haugus*, loosely translated means "vacuum" and is the name

---

1. Helfman and Randall (1973) were able to obtain the derivations of only about 40 of the 336 Palauan fish names they listed. I was not able to expand appreciably on this.

Robert Randall (pers. comm.) has noted that a surprisingly high percentage of Tobian fish names appear to have remembered derivations and that this would not be expected in a culture familiar with fish for hundreds of years. Possibly some of the derivations obtained from my informants were erroneous. Indeed it may well often be impossible for speakers of any language without a written history to distinguish between correct and merely plausible word derivations, unless he or she is privy to the recondite procedures of the lexicostatistician. But in the present study the linguistic legitimacy of a fish name derivation was of less importance than its descriptive function. As long as the latter reflected accurately some characteristic of the fish in question, it served the purpose at hand.

used for any large grouper; the ability of groupers to "inhale" their prey by means of the sudden expansion of the oral cavity is well known. *Martacham* means "very smart fellow" and refers to a squirrelfish which reputedly will not bite a second time if it feels the prick of the hook. A small deepwater emperor will not bite unless the bait is placed in close proximity to it. Its name, *watur*, means "never moves." *Hafira* means "testing" and refers to the cautious nibbling approach of a certain snapper to a baited hook. (A snapper with similar habits, *Lutjanus semicinctus*, is called by Palauans *mengeslbad* meaning "jokes when biting.") *Rau*, meaning "study," is the name of a pomodasyid, *Plectorhynchus chaetodontoides*, which also is circumspect in its approach to bait.

Other names refer to the fighting characteristics of certain fish. *Teter* is the name of a jack (carangid) which tends to run toward the fisherman when hooked. The name refers to the action of pulling in slack fishing line. *Hao* means "missing" and is the name for two species of parrotfish, *Cetoscarus pulchellus* and *Chlorurus strongly-cephalus*, that are often able to shake the hook. *Fotorimar* means "like a man" and refers to the strong fight put up by the small sea perch, *Anthias huchti*, when hooked.

Other Tobian fish names refer to the habitats in which these fish are found. The generic name for squirrelfish, *mor*, means "in mourning" and refers to the retiring habits of these fish, which typically remain in caves during the day. Stonefish and related scorpaenids are called *reyu*, meaning "lazy," referring to their habit of lying motionless on the bottom. The generic name for goatfish is *so'owo*, which means "middle of the current." According to Tobians several species of goatfish, when in schools, favor strong currents. (I have not noticed such tendencies, but Harry [1953] notes that goatfishes at Raroia congregated where currents were particularly strong, and Helfman [pers. comm.] notes that schools of the mullet, *Mulloidichthys flavolineatus*, in Hawaii often favor strong currents.)

The number of names of fish that reflect their association with drifting logs attests to the importance of these logs to South West Island fishermen. The name *usurifarema* is derived from words meaning "slow" and "drifting log" and refers to a sluggish frogfish sometimes found associated with drifting logs. *Faumer* similarly means "drifting" and is the name of the tripletail, *Lobotes surinam-ensis*, which is often found hanging motionless on its side under a floating log and looking very much like a drifting leaf. *Ruhuruho* is an old generic name for triggerfish, now seldom used. It means "cast net." It refers to the fact that the young of certain triggerfishes are found clustered around drifting logs, where they are caught with cast nets and used for tuna bait.

Tobians suspect that the moorish idol, *Zanclus cornutus*, does not spawn locally because juveniles are never seen. They believe that these fish come from elsewhere accompanying drifting logs. Hence the name *wasori*, or "foreigner." On Sonsorol most small dolphinfish are caught in the vicinity of floating logs. The Sonsorolese name for small dolphinfish is *riwesiri yapetase*, meaning "children from the country of the white people." The logs which drift past Sonsorol usually come from the west, moving on the Equatorial Counter Current. The first white people to set foot on Sonsorol also sailed into view from the west, giving the Sonsorolese the impression that their homeland must lie to the west.

The names of other fishes derive from their distinctive habits. *Chera* is the name applied to a woman who dotes on a man and will not let him out of her sight. The name is also applied to remoras and alludes to their habit of following or actually adhering by means of an adhesive disc on their head to larger fish and turtles. Another fish that is well known for its habit of accompanying a larger fish is the pilot fish, *Naucrates ductor*. Its name in Tobian, *yetam*, means "travels with a father."

The nurse shark, *Ginglymostoma ferrugineum*, is called *sabacho*. The name is derived from two words meaning "sleep" and "awkward" and refers to the ungainly swimming motion of this fish and its habit of resting motionless for long periods on the bottom. *Urech*, the name for *Heniochus acuminatus*, a false moorish idol, means "erratic" and refers to the jerky, tortuous path taken by this fish when it is pursued.

The imperial angelfish, *Pomacanthus imperator*, makes a loud popping noise when alarmed. Its name, *Ngungpaha* means "son of thunder." *Ngusngus* is an onomatopoetic name describing the alarm sound emitted by a species of squirrelfish. Similarly *bucho*, the Tobian name for the triggerfish, *Balistapus undulatus*, describes the sound made by this fish when disturbed.

Yassur, one of the Tobian names for dogtooth tuna, means "discourage" and refers to its reputed habit of chasing away other species on the fishing grounds. *Suchowa* and *merifuts* are names for two species of barracuda. Both words mean "dangerous."

There is a wrasse on Tobi reefs that ranges over large areas of the reef, often "getting into trouble" with other species whose territories it invades. According to Patris its Tobian name, *tanganangan*, means "delinquent."

The diets of some fishes are reflected in their names. *Richopah* is the name of a damselfish that inhabits shallow water where Tobians used to relieve themselves. Its name means "little excrement fish," for its diet includes human feces. The cleaner wrasse,

Labroides dimidiatus, eats the external and mouth parasites of larger fishes which hover motionless in the water as if mesmerized, allowing themselves to be cleaned. The Tobian name for this fish is *chariferigut,* which comes from two words meaning "soothe" or "hypnotize" and "suck." (Such cleaning behavior was discovered by biologists only about twenty years ago.) The giant bumphead parrot-fish, *Bolbometopon muricatus,* is called *matirai,* meaning "pregnant." The name refers to its oft-protruding stomach, which, on Tobi, is often found to be stuffed with sea urchins. *Cherouchouko* means "chum together" and is the name given to certain jacks. The name refers to the custom of throwing large quantities of bait into the water to attract schools of these fish and keep them in the vicinity for a day or more.

Other species of fish are named on the basis of some physical peculiarity. *Bub,* a word roughly equivalent in meaning to "parallelogram" in Tobian, is in widespread use in Micronesia, including Tobi, as both the family name for triggerfish and the constellation we call the Southern Cross, both of which approximate a parallelogram in outline. Ostraciontids, descriptively named "boxfish" in English because of their hard, box-like carapace, are similarly named *tavitef*—fishing box—in Tobian. (The Tobian fisherman keeps his hooks and lures in a waterproof box made from breadfruit wood.) The angelfish, *Centropyge bicolor,* is called *hob* or "pandanus fruit" because its rough scales are reminiscent of the texture of this drupe.

Marlins are called *tauchacha.* The name also refers to a Tobian weaving device constructed from a pair of marlin bills. The sailfish is called *moharechoh,* a name that means "decorative mat" and refers to the fish's large dorsal fin.

Butaha literally means "coconut vessel" and refers to the use on Tobi of a half coconut shell as a bowl. The loose translation or implication of this name when applied to certain filefish means "liver so big it fills the bowl." These species have large and esteemed livers.

The names of some fishes relate to their smell or taste. Sharks are *paho* meaning "[smelling of] excrement." Their flesh tends to smell of ammonia. Another fish (probably a grammistid or soapfish) is called *temaubour,* meaning "smells bad." *Teribour,* the name of a pomadaysiid, means "smell stinks." The sailfin tang, *Zebrasoma veliferum,* is called *pingao,* which means "special smell." The bumphead wrasse, *Cheilinus undulatus,* is called *mam,* meaning "fat." This fish is esteemed for its rich, oily flesh.

Other fish names reflect certain South West Island customs. *Uremar* means "to become a man." It is the generic name given to several species of pomadaysiids whose black markings resemble the tattoos that used to be applied to Tobian males on their reaching adulthood. *Pygoplites diacanthus*, a blue-banded angelfish, has markings on it similar to those seen tattooed on the elderly natives of another South West Island, Pulo Anna. Its name, *farupon pisahe he pulo*, refers to that similarity.

Needlefish are too long to be conveniently packed in fish baskets for transport so they are generally strung on a cord or vine like beads on a necklace before being carried up from the beach. The generic name for needlefish is *mahi*—"necklace." *Paip* is the name given to hollow bone straw through which infirm Tobians used to suck water from coconuts. It is also the generic name for butterfly fish and refers to the protruding tubular mouths of many species.

A common illness on Tobi is characterized by chancre sores and fever. The name of this illness is *moghu*. It is also the name for a surgeonfish, *Acanthurus glaucopareius*, used in the treatment of this malady. The fish is ground up without removing the internal organs and eaten. *Horach* is the name of a tree from which another medicine for treating *moghu* is made. It is also the name of another surgeonfish, *Acanthurus triostegus*, used to treat *moghu*.

Starting with Durkheim and Mauss (1903) anthropologists have examined folk taxonomies mainly as a means of deducing some of the thought processes of the peoples who devised them (e.g., Berlin et al., 1974; Bulmer et al., 1975). Hunn, for example, maintains that the "ultimate goal" of studying folk taxonomy is "to construct a theory that adequately accounts for the pan-human ability to form concepts and to organize them into efficient systems" (Hunn, 1977, p. 5). But the foregoing discussion demonstrates that focusing on the derivations of folk taxonomic names can provide other types of useful information, yielding new insights concerning the things named, their significance to the natives, or other aspects of native culture.

1. Some of Palau's Rock Islands. Chapter 1

2. Ngiraklang. Chapter 1

3. Making a leaf sweep. Chapter 2

4. Gathering in the leaf sweep. Chapter 2

5. Young helpers spearing fish trapped in the *kesokes* net. Chapter 2

6. A Palauan spearfisherman preparing to submerge feet first. Chapter 2

7. In pursuit of milkfish. Chapter 2

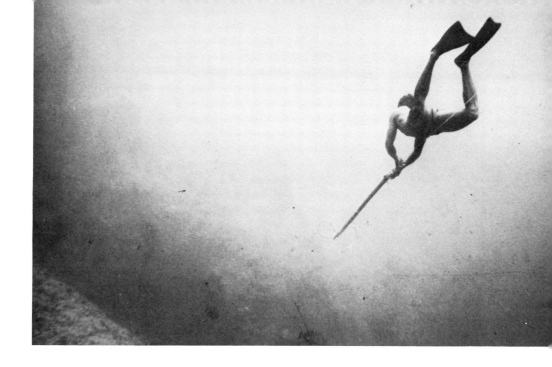

8. Palauan spearfisherman an instant after firing spear. Note the speared surgeonfish at the bottom of the picture. The diver holds the butt of the gun with both hands so as to cushion the substantial recoil of the gun. Chapter 2

9. Spearfisherman waiting at drop-off for fish to come to him. Chapter 2

10. Palauan throwing *bidekill*. Chapter 2. Photo by Karen Strauss.

11. Ngiraklang making a traditional Palauan fish trap. Chapter 2

12. Palauan fish trap. Chapter 2

13. Palauan coaches his young son. Chapter 2. Photo by Karen Strauss.

14. Spawning aggregation of blue-lined sea bream. Chapter 3

15. Spearfishermen with catch of blue-line sea bream from the Peleliu spawning aggregation. Chapter 3

16. Tobi. Chapter 7

17. Patricio making breadfruit leaf-fishing kite. Chapter 7

18. Patricio about to release fishing kite. Chapter 7

19. Patricio and his grandson using fishing kite. Chapter 7

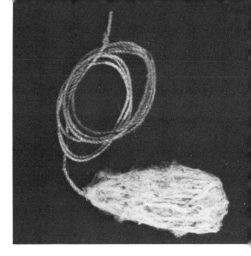

20. Spider web fishing lure. Chapter 7

21. Old Tobian fishhook. The irregular grooves on the hook surface indicate that it was shaped using a piece of coral as a file. Chapter 9

22. Kalisto demonstrating the device for cutting fishhook blanks. Chapter 9

23. Tip of Tobian device for cutting fishhook blanks out of turtle shell. The shark tooth on the left is the pivot point, the two teeth on the left are the cutting points. Chapter 9

25. Piece of Tobian sennet fishing line used for five generations. Chapter 9

24. Zacharias making a fishhook from steel wire using a rock as an anvil. Chapter 9

26. Palauan cleaning trochus shell. Photo by Karen Strauss.

CHAPTER

THE ARBOREAL OCTOPUS

11

There are many examples of marine animals with habits that once seemed fantastic but are now well documented: anglerfishes which attract their prey with built-in fishing lures; flounders which emit a substance that paralyzes the closing jaws of an attacking shark; eels and rays that possess living storage batteries with which they deliver powerful shocks; gobies that climb trees; fishes whose spawning coincides precisely with the phase of the moon. Biologists accept these stories today because they have all been scientifically documented.

But intense skepticism greets accounts of such strange phenomena until they have been confirmed by one or more reputable scientists. The claim, for example, that archerfish catch insects by shooting them out of the air with accurately aimed jets of water seemed so improbable that biologists refused to believe it for 138 years after it was first reported (Gill, 1909). (Archerfish can commonly be seen today in shallow estuaries in Palau, performing like animated water pistols.)

Skepticism is essential to science. For every improbable story that turns out to be true there are many that are not, and it is the job of the scientist to sort one type from the other. In areas such as

Micronesia which abound with little-known species it can be taken for granted that some have surprising habits yet to be discovered or verified[1] by biologists, though well known to local inhabitants. I was given a number of accounts of such habits which I did not witness personally. Some of them stretch the imagination. Only two of them, as far as I know, have been witnessed by a trained biologist (see below). I was able to disprove one. But, as will be seen below, it was based on a reasonable interpretation of available facts. Judgment must be withheld on the rest of these accounts until more relevant facts and observations come to light. But I feel confident that some of them will ultimately be verified.

I am convinced that the fishermen who volunteered these reports believed them to be correct—believed that they had witnessed the behavior they described. Earlier I described the general reliability of these men as informants. I believe it would be a disservice to them to ignore these particular observations. It would also be a disservice to science; those that are correct are much less likely to be verified scientifically if no one brings them to the attention of the scientific community. Most were related to me independently by at least three informants whose stories did not differ significantly.

How the Octopus Fights Back

Several Palauan fishermen told of having witnessed a novel form of defense employed by an octopus when attacked by groupers, eels, sharks, or triggerfish. The octopus is said to jump onto its attacker, enveloping the latter's head with its body and tentacles. Eyes covered by the octopus, the fish whirls and shakes trying to dislodge it. The octopus suddenly lets go and, taking advantage of its attacker's disorientation, jets rapidly away to shelter. Trukese, Ponapean, Marshallese, and Gilbertese fishermen volunteered similar observations.

I have never read any account, scientific or otherwise, of such behavior, and such a story might well evoke skepticism. But a film made by Ben Cropp, an Australian underwater photographer, lends considerable support to it. While Cropp was photographing an octopus on the Great Barrier Reef, a shark swam into the picture and attacked the octopus. The octopus clung tightly to the shark's head. The shark thrashed around violently for a few seconds, at which

---

1. It is, of course, somewhat misleading to use the word "verify" in this context. The tenth-century Palauan who watched archerfish knock insects out of the air since the day he was old enough to toddle down to the tidal creek that ran through his village would be amused if he knew that a thousand years later a foreigner (a European scientist) would claim credit for having "verified" this phenomenon.

point the octopus suddenly released its grip and sped off to safety. The entire sequence was recorded on film, in focus, at close range.

As anyone knows who has ever grabbed an octopus, the animal usually does not retreat, but rather throws its tentacles quickly around one's hand and arm. By the same token, if a natural predator gets a portion of the octopus in its mouth, the reaction of the octopus would probably be similar—that is to throw its body and tentacles around the closest portion of the attacker: the head. Once the predator is sufficiently disoriented by this action to open its mouth and relax its grip (much as a startled neophyte human octopus catcher instinctively lets go when he feels his arm firmly gripped by a bunch of long, sticky tentacles), the likely response of the octopus would be to let go and try to escape. The action in Cropp's film is too fast to determine whether the octopus "jumped" purposefully onto the shark's head or simply clung to its head after having been gripped in its jaws.

Pelagic Fish Orient to Ocean Currents

Early in this century fish behaviorists discovered that some fishes orient visually to currents. When stripes were painted across the bottom of an experimental flume providing visual clues to the presence of currents, the fish swam facing into the current. But over a featureless bottom they drifted with the current. The fish used in these experiments were shallow-water species that are normally in visual contact with the bottom. Open-ocean pelagic fishes, in contrast, spend their lives out of sight of the bottom. They are like men in a submarine—without obvious visual clues to their motion relative to the earth's surface. Have such fish developed nonvisual methods of detecting their movement? Many investigators doubt it. Harden-Jones (1968, p. 16), for example, states: "Fish have never been shown to react to water flowing at a constant linear velocity unless they are close to or in sight of the bottom."

Ngiraklang, however, told me that schools of tuna feeding over deep water off Ngeremlengui generally move into the current because the schools of small fish which they feed usually head into the current. Dick Shaver, manager of the Van Camp tuna plant in Palau and an ex-tuna fisherman, emphatically endorsed Ngiraklang's observations. Commercial fishermen in Hawaii say that it is common knowledge among them that various kinds of oceanic fishes usually travel heading into the prevailing current. Japanese fishermen have made similar observations (Kishinouye, 1923).

A fisherman is able to detect the direction of the current when he is within sight of land by observing the drift of his boat relative to the land. But light refraction at the air−water interface prevents fish from seeing a reference point so low on the horizon. It would appear then that some pelagic fish can detect the direction of flow of the water mass in which they are swimming in the absence of visual clues. How is this possible?

One possible explanation relates to the fact that a water mass moving through the earth's magnetic field produces a weak electrical field (Longuet-Higgins et al., 1954). The field is directional and has polarity. A number of fish have recently been shown to be able to detect such fields (e.g., Akoev et al., 1976; Kalmijn, 1974, 1978.) Whether they use them in their migrations—and whether this explains the observations of Ngiraklang and the other fishermen—remains to be established.

Bleary-Eyed Reef Fish

Palauans and other Pacific island fishermen maintain that it is easier to get fish with an underwater speargun around dawn and dusk than during the middle of the day. I asked several Palauans what the explanation might be for this decreased wariness. They replied that the fish really did not seem less cautious, they simply seemed "bleary-eyed"—unable to see very well at these times. Aristotle similarly noted that "fish are weaksighted near dawn and dusk."[2]

Many reef fish feed at night and are able to perceive their prey at much lower light intensities than they encounter around dusk and dawn. Why then should such fish have difficulty in perceiving predators in twilight? A possible answer comes from laboratory studies of fish vision. The vertebrate eye contains two sets of photoreceptors: rods and cones. Rods mediate vision at high light intensities (photopic vision). Cones take over this function at lower light levels (scotopic vision). When the light level changes from high to low, or low to high, a period of accommodation occurs during which time rods gradually take over the main visual function from cones or vice versa. Thus we experience a brief period of near blindness when we enter a movie theatre and again when we exit.

According to Munz and McFarland (1973), "Twilight is the most difficult period in which to perform visual tasks—witness the problems of a human driver during evening rush hour, while he is poised between the threshold for photopic vision and the upper limit for scotopic vision. *Twilight is even a more difficult time for fishes, since their light and dark adaptation are slow processes and require hours rather than minutes*" (emphasis added). Munz and McFarland's conclusion is based on histological examinations of the

---

2. Aristotle spent two years on the island of Lesbos. Much of what he subsequently wrote about marine life in *Historia Animalium* he learned from the fishermen of the island.

retinas of fishes exposed to different light intensities. Do the observations of Aristotle and Palauan fishermen provide a behavioral corollary?

A Shark That Vacuums Its Food

A Tobian fisherman reported having seen a nurse shark on Helen Reef extracting a small giant clam from its shell by suction. Such an act would require the generation of considerable vacuum pressure, for giant clams are strongly secured to their shells by their large adductor muscles. I dismissed the observation at the time, guessing that what had been seen and misinterpreted was simply the removal of a small temporary resident, perhaps and octopus, from an otherwise empty shell.

Recent research, however, lends weight to the fisherman's account. Moss (1965) observed that nurse sharks possess skeletal and muscular adaptations that suggest that they can feed by means of a suction-pump mechanism. To test this hypothesis Tanaka (1973) designed a special feeding chamber. In order to obtain food from within it, nurse sharks had to cover an opening in the chamber with their mouths and apply suction. The vacuum pressure they generated was measured by means of the displacement of a plunger in the chamber. Tanaka reports that nurse sharks attempting to extract food from the chamber positioned themselves vertically over it, applied their fleshy lips to the hole so as to seal the opening, then inhaled. The pull thus generated by 1.5 to 2.1 meter nurse sharks amounted to as much as 44 kilograms (96 pounds).

Interestingly the Palauan name for the nurse shark, *metmut*, comes from the verb "to suck" (Helfman and Randall, 1973).

Fish Blood as a Fish Tranquilizer

When a leaf sweep has been drawn in and the fish are tightly grouped within it, Palauan fishermen gather around its periphery and prepare to remove the fish one by one by spearing them. This is a crucial time, for if the fish are disturbed too much by the splashing and movement around them, they dash about wildly and, like stampeding cattle, are liable to leap over or crash through the flimsy barrier.

Palauans have discovered that if they gingerly spear one large fish from the school, cut its gills, and throw it back into the school, the bleeding fish will swim around trailing blood. And when this happens the other fish tend to crowd together, drop to the bottom, and become almost inactive. The likelihood of a "stampede" is thus greatly reduced. I did not have an opportunity to see this technique used (the leaf sweep has been all but superseded by the *kesokes* net in the last generation), but was assured by Ngiraklang and others that it worked well.

At Ant Atoll near Ponape in the eastern Caroline Islands the palm frond scare-line is used to catch schools of kawa kawa (*Euthynnus affinis*) which occasionally enter the lagoon and swim along the edge in shallow water. Here, too, I was told by Ponapean fishermen, the blood of the first fish to be caught is dripped into the water to calm the rest of the school.

Many commercial tuna fishermen believe that if blood from landed tuna washes off the deck and into the water among feeding tuna, it will cause them to stop biting. Does blood really affect reef fish and tuna, and if so, by what means?[3]

Strasburg and Yuen (1958) observed that blood poured into sea water in concentrations that turned the water red but did not significantly reduce its transparency had no apparent effect on feeding skipjack tuna. When they poured food coloring into the water, however, the tuna avoided opaque patches of dye without exception but moved into the dyed water to feed once the dye was sufficiently diluted so that the water was once again transparent. Thus blood, as such, apparently did not inhibit these tuna, but a dissolved substance that reduces visibility significantly did affect them.

Although Palauans choose the largest fish they can spear to spread blood among the rest, they maintain that the "calming" effect they observe cannot be caused by obscuring of the fish's vision; the effect is observed when the water is still reasonably transparent. However, shallow water that appears only slightly turbid when one is standing in it looking down through perhaps a foot of it nonetheless often appears quite murky when one submerges and looks off horizontally through it. In other words, what appears only slightly cloudy to fishermen looking down into the water may appear quite turbid to a fish. Moreover, blood would cause a red shift in ambient light. Reef fish have retinal pigments that are poorly developed in the red-absorbing range, and the tuna, *Euthynnus affinis*, has no red-absorbing pigment at all (Munz and McFarland, 1975). Consequently this red shift may hamper vision to a significantly greater degree than the decrease in total light produced by the blood in the water. In short, blood may affect these fish simply by limiting visibility.

3. It might be surmised that this behavior is caused by something other than blood released by an injured fish. It is well known that some fishes when disturbed exude an "alarm substance" from their skin which communicates the presence of danger to other fish. Tests of a wide variety of reef fish for the ability to release such substances were all negative (Pfeiffer, 1963), however, as were tests with tuna (Strasburg and Yuen, 1958).

A second line of evidence to support this explanation is provided by the behavior of *Euthynnus affinis* in captivity when exposed to clouds of suspended clay. During initial trials these fish schooled tightly and slowed down almost to a stop (Barry, 1978)—behavior very much like that which reportedly occurs when this same species encounters a cloud of blood. Hobson (1972a) and Stevenson (1972) have also observed that small plankton-feeding reef fish descend close to the reef when turbid water moves into the area, possibly in response to the increased threat from predators occasioned by decreased visibility.

Poison-Breathing Sea Snakes

The *mengerenger* or banded sea snake, *Laticauda colubrina*, is a common resident of Palau's reefs, often seen in shallow reef communities. Typically it swims slowly along, its darting tongue frequently tasting the water just like a land snake tastes the air. It spends many hours searching in holes and crevices for the small fish on which it feeds.

The *mengerenger* possesses a venom that is several times as potent as that of the king cobra. But it very rarely strikes humans, even when roughly handled. Moreover there are few portions of the human anatomy into which it can sink its fangs; they are small and located in the rear of its small mouth. Palauans were thus unaware of its poisonous bite until they heard about it from Westerners. In fact some adventuresome Palauan children still play with this snake. (In Fiji, where the species is very common, guides sometimes actually encourage adventuresome tourists to wrap these snakes around their necks.)

Palauans were aware that there was *something* poisonous about it, however. They noticed that sometimes when the *mengerenger* withdrew its head from a hole where it had been searching for food, small fish would come gyrating out in obvious distress and swim convulsively for a few seconds before being caught and swallowed by the snake. This led to the belief that the *mengerenger* "breathes poison" into the water. Before the advent of face masks divers were cautioned to close their eyes if they happened on such a scene for fear that the poison in the water might blind them.

To test the possibility that the venom could affect fish merely by coming into contact with them in the water, Dr. Bill Dunson and I milked the venom from a large *mengerenger*. We poured the fresh venom—more than enough to kill a man—into a one-liter container with several small reef fishes. None showed any signs of distress and were still alive several hours later. Thus it appears that the *mengerenger* bites small fishes it traps in holes in the reef, injects its venom, then withdraws and waits for the fish to succumb before capturing it. (Similarly Klemmer [1967] observed that a related species, *Laticauda laticaudata*, lets go after biting a fish and waits for it to die before swallowing it.)

Because Palauans had never seen one of these animals inflict a poisonous bite (the snake generally bites while its head is hidden in a hole where it has cornered a fish), and because there are no aggressive poisonous land snakes in their islands, Palauans were unfamiliar with the ability of some snakes to inject venom. Under the circumstances their "poisonous breath" explanation for what they observed can be seen to be quite plausible.

The Cornetfish and the Moray Eel

Cornetfish are rather rigid, elongate fish that resemble mobile sticks as they propel themselves through the water using small, almost transparent fins. They are common in coral reef areas throughout the tropics. Many Palauan fishermen I interviewed reported having seen at least once while diving a remarkable interaction between cornetfish, *Fistularia petimba (ulach)*, and medium-sized moray eels (*kesebekuu*). The *ulach* is said to approach an eel whose head protrudes from a hole in the reef and suddenly ram its snout down the eel's throat. (Morays typically rest with their heads sticking out of holes and their mouths open.) The *ulach* twists around for a few seconds and then withdraws, its snout covered with fragments of flesh. The eel is said not only to seem helpless while suffering this indignity but also to appear quite debilitated, sometimes to the point of being unable to withdraw into its hole after the ordeal is over.

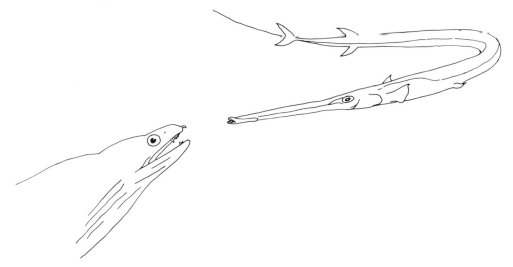

After hearing several eyewitness accounts of this behavior from Palauan fishermen I had an opportunity to find out if fishermen elsewhere had witnessed similar behavior. While in Yap, an island 300 miles from Palau inhabited by people of a quite different culture and language, I asked five different fishermen whether they had ever seen a cornetfish and a moray eel do something strange together (purposely not giving any further details as to what I was looking for). All five immediately volunteered that they had witnessed cornetfish ramming their snouts down the throats of moray eels. Yapese accounts differed from most Palauan accounts in only one way. The Yapese surmised that the cornetfish was eating the intestines of the eel, whereas most Palauans said they believed that the cornet fish was "vacuuming" out the stomach contents of the eel.

I recounted these stories to Mike McCoy, an American fisheries officer on Yap. McCoy is married to the daughter of Relugmai, a chief of the Satawal Islanders, a third cultural group in the Caroline Islands. Several months later, McCoy described in a letter a subsequent conversation he had with Relugmai:

McCoy: "He [Johannes] told me about this one kind of fish, a cornetfish, and what it does to catch eels."

Relugmai: "Wait. Do you know how they catch eels? I've seen them catch them."

McCoy: "No, how is it done?"

Relugmai: "They go inside the eel's [sao-fung] mouth, then twist around inside. The surface of their 'beak' is abrasive, and this cuts up the eel's insides. Then they back out again and the eel dies."

McCoy: "Have you actually seen this?"

Relugmai: "Yes, at West Fayu [a nearby island] I've seen them do it, but not to really big eels, only the smaller ones" [putting forefinger and thumb together in a circle].

Relugmai's description of the "beak" or snout of the cornetfish as being abrasive is correct. The eyes of the cornetfish are set back several inches behind a long, rough, almost fleshless snout. The snout is angled in cross section and each angle is rather sharp. When a live cornetfish is picked up by the snout it twists in the captor's hand, and the effect is often sufficiently painful so that one either drops the fish or quickly shifts one's grip to behind the eyes.

Concerning Relugmai's comment that the eel dies after this encounter: most fishermen from Yap and Palau who said they witnessed this phenomenon while diving simply said that the eels seem exhausted afterwards. But one Yapese fisherman said he saw this happen in a fish trap and that within a short time the eel was dead. And one Palauan fisherman who, after witnessing this event, speared the eel in the body and tossed it into his boat volunteered that the eel died with atypical speed. (Moray eels typically endure extensive injuries plus long exposure to air before expiring. They are among the most tenacious of life of all reef fishes.)

The advantages of such behavior to the cornetfish is obvious if it derives food (guts or gut contents) from the encounter. But why should the moray eel tolerate such an assault? Why doesn't it simply withdraw into its hole, thereby forcing the cornetfish out of its throat? A possible explanation comes to mind.

Some fish are paralyzed when touched in certain spots. For example, if one is removing a hook from the mouth of a North American largemouth bass, one need only grasp the fish by the lower jaw, inserting the thumb in the base of the mouth behind the lower front teeth, in order to cause the fish to go limp instantly. Palauans told me of having observed similar pressure points on other species. The squirrelfish, *Adioryx spinifer* (*desachel*), and the scad, *Trachurus boops* (*terekrik*), can also be calmed by grasping the lower jaw between thumb and forefinger. The lutjanid, *Aphareus rutilans* (*metngui*), and the lethrinid, *Lethrinus ramak (udech)*, will cease struggling when their first dorsal fin is depressed. Perhaps moray eels have such a pressure point within their mouths or throat and the hard bony snout of the cornetfish activates it.

I had heard one third-hand account of a Sonsorol man who was said to have caught small moray eels by grasping them, palm up, in such a way that one finger was inserted in the mouth, reportedly paralyzing the eel. Later in Hawaii, wearing pieces of thick rubber tubing to protect my fingers, I attempted to emulate him using moray eels dipnetted from a laboratory holding tank. I found that if I firmly squeezed the roof of the mouth between my inserted finger and my thumb, struggling eels usually went completely limp. They remained dazed and motionless for a moment when released and returned to the tank. As my thirteen-year-old assistant with these experiments, Bren Caperon, pointed out to me, a cornetfish expands its snout when feeding, and this action might be a means by which it could exert a similar upward pressure on the roof of an eel's mouth.

The trick does not work with all species of morays; subsequently I was unable to paralyze in this fashion an unidentified species of moray eel in southwest Australia.

Another "Archerfish"

A well-known member of the snapper family in Palau called *kedesauliengel* (*Lutjanus argentimaculatus*) is found in a wide variety of habitats but most typically frequents mangrove swamps and tidal creeks. According to Palauans this fish employs at different times two remarkable techniques for obtaining food. Like the archer-fish, *Toxotes jaculatrix (uloi)*, it reportedly captures prey by using a jet of water ejected from the mouth. The archerfish has a special groove in the roof of its mouth from which, when its tongue is forced upward, a stream of water is ejected. *Kedesauliengel* have no such groove. This is not inconsistent with fishermen's descriptions of the spitting pattern of this fish. Whereas archerfish eject a long thin stream of water, mangrove snappers eject a broad gush of water from their mouths. Although the range of the two species is similar, the volume of water ejected by snappers is much greater, and so, reportedly, is the size of their prey.

Mangrove rats, snakes, and lizards are all knocked off overhanging mangrove branches by *kedesauliengel*, it is said. If the fish fails on its first attempt to dislodge its prey, it will shoot repeatedly and, more often than not, finally achieve its purpose. Whereas its accuracy is reputedly admirable, its choice of targets would appear to be fallible. Occasionally, according to fishermen, millipedes, *Polyconoceras allosus (choas)*, are knocked down and eaten, resulting in the death of the fish due to the effects of the millipede's hydrocyanic acid venom. In addition one informant reported having been subjected, when he was using an overwater latrine, to an unsolicited caudal lavage by a misdirected *kedesauliengel*.

Yapese and Ponapean fishermen independently related similar accounts of the spitting behavior of this species but I was unable to observe it. The fish is highly valued as food. Consequently its numbers in the vicinity of the heavily fished tidal creeks where I lived were low and I seldom saw the fish at all.

According to a considerable number of reports by both Palauan and Ponapean fishermen, the mangrove snapper also uses its tail to capture terrestrial prey. The fish is said to lie with its tail resting on the bank of a tidal creek or on a half-submerged log. The tail is moved up and down. The movement or the sound produced by it (which is reportedly often quite audible) attracts rats or crabs. When such an animal touches the tail, the fish suddenly sweeps it into the water and eats it. Most informants also volunteered that flies were often attracted to the tail, perhaps by the strong-smelling mucus, and that lizards were sometimes attracted to the flies and also knocked into the water and eaten.

To my knowledge such behavior in snappers has never been described by a scientist. But another Palauan fish, a freshwater eel, *Anguilla marmorata, kitlel*, has been observed by biologists performing a similar feat. I first heard about it when a Peace Corps biologist, Daryl Gray, told me of observing what appeared to be a dead, dried-out looking eel lying on the top of a log protruding from a small stream near his house in Ngiwal. Wondering how the eel could have come to be deposited above water level, he kicked it into the water and was surprised to see it swim away.

Palauans subsequently told me that the *kitlel* commonly comes out of the water and lies on the bank, often attracting flies. A rat will frequently come to investigate the eel which then bites the rat and knocks it or tugs it into the water. The eel then re-enters the water and swallows the rat. This behavior has been witnessed by Trust Territory Entomologist, Mr. Demei Otobed.

Fish Mortalities Induced by Light Rain

It is commonly stated by Palauans that when a short rain occurs on a hot, still afternoon around low tide it is sometimes followed within a few minutes by the death of large numbers of small fishes on shallow reef flats in the lagoon. The phenomenon, referred to in Palauan as *delides*, reportedly occurs most commonly near Peleliu. About ten years ago fish biologist Gene Helfman observed a catch of atypically small fish, said by Peleliuan fishermen to have been harvested after one of these sudden mortalities.

Rampaging Sharks

There is one circumstance that Palauans often describe in which a diver will leave the water quickly for fear of sharks even before seeing them. On such occasions packs of sharks, often containing more than one species, swim very rapidly along the outer reef edge. Their approach may be heard before they are seen by means of a scraping sound possibly made, according to divers, by the rubbing together of the abrasive skin on the tails of the sharks as they swim in tight formation. The phenomenon is known in Palauan as *plutek*. Eibl Eibesfeldt (1964) similarly observed that a diver "can hear big sharks coming when they are in a hurry." Sharks exhibiting this behavior are said to be exceptionally aggressive. Small blacktip and whitetip reef sharks reportedly leave the area immediately when such a sound is heard. I have asked a number of experienced American divers, including two shark behavior specialists, about this phenomenon but none had ever encountered it.

"Mobbing" of Sharks

Eight different Palauan fishermen told me independently of having seen jacks (carangids) attack sharks, and that when this happened the jack was not only always the aggressor but also invariably the "winner" in the encounter. This behavior reportedly occurs in several species of jacks including *Caranx melampygus (oruidl)*, C.

ignobilis (cherobk), and *Gnathanodon speciosus (wii)*. Usually a single jack is involved, but on some occasions two or more jacks will reportedly "gang up" on a shark.

The attacker butts the shark repeatedly in the flank. Each time the shark flinches, then attempts to bite the jack. The jack evades this counterattack by darting underneath the shark's belly and maneuvers to stay out of sight of the shark no matter how the latter twists and turns. (When two jacks are involved one will often position itself directly above the shark between attacks.) After a few seconds the jack will move out quickly and ram the shark again. Sometimes the force of these blows on the shark's abrasive skin is sufficient to lacerate the head of the attacker, causing it to bleed. The beleaguered shark is powerless to defend itself against the more agile jack and its strength gradually wanes. After repeated attacks the shark's gills are sometimes seen to stream blood, and several fishermen said they saw the shark ultimately sink to the bottom, apparently mortally wounded. Marshallese fishermen also told me of having seen the jack, *Caranx melampygus*, butt sharks repeatedly.

What purpose would such an attack serve? I do not know, nor did the fishermen. But there are some well-known parallels, when the hunted becomes the hunter, among land animals. Groups of blackbirds, for example, will attack and harass owls and other birds of prey, and groups of California ground squirrels will attack snakes (e.g., Owings and Coss, 1977). This phenomenon has been named "mobbing." Its purpose remains unresolved. Mobbing apparently always involves prey attacking larger and stronger potential predators, but it generally seems to occur at times when the predator poses no immediate threat. (In most cases the predator is nocturnal and the assailants diurnal.) When it is the predator which initiates an attack, flight is more often the response of the prey.

This is not the first report of mobbing in marine animals or, for that matter, coral reef animals. Eibl-Eibesfeldt (1964) reports observing fusiliers mobbing moray eels in the Indian Ocean, and Hobson (pers. comm.) has seen a number of reef fish mob moray eels in Hawaii. In addition somewhat similar behavior was noted by Limbaugh (1963) in a rainbow runner (a relative of the jacks mentioned above) which "bumped a small shark causing it to disgorge some food which the more agile runner promptly swallowed." T. Clarke (pers. comm.) observed an encounter between a jack and a shark similar to that described by Limbaugh except that the shark was hooked and fighting the line at the time. Eibl-Eibesfeldt (1964) photographed schools of rainbow runners scraping themselves on sharks, perhaps, he speculates, using the shark's abrasive skin to remove parasites.

In short, although there are no scientific records of mobbing of sharks by jacks, related observations lend weight to Palauan accounts, although the purpose of such behavior is a puzzle.

How Some Sharks Give Birth

Little is known about the process of birth in most sharks, although it is known that females of a number of species come into shallow water to release their young (e.g., Clarke, 1971; Springer, 1960). Ngiraklang and several other Palauan fishermen told me that they had occasionally witnessed sharks giving birth. Their observations were usually made near dawn or dusk at high tide (which places them near the time of new or full moon) at the seaward edge of mangrove swamps. The sharks were said to "look for" a mangrove branch hanging into the water or a sunken log. They would then rub their bellies vigorously against this object as they swam past. After each pass a single pup was released. Several species of sharks were mentioned, but the ones most frequently noted were the reef blacktip, *Carcharhinus melanopterus (matukeyoll)*, and the grey reef shark, *Carcharhinus amblyrhynchos (mederart)*.

These accounts suggest that some species of sharks may experience difficulty in expelling their young and employ vigorous body movement to aid in the process. Observations on two other species of cartilaginous fish support this notion. Wass (1973) reported that a pupping sandbar shark made sudden turns while swimming rapidly, releasing one pup at the pivot point of each turn. One Palauan fisherman also reported having seen eagle rays, *Aetobatus narinari (ochaieu)*, give birth to young while jumping from the water. The same behavior has been reported from the Society Islands (Bagnis et al., 1972), the Seychelles (Ommaney, 1966), and North Carolina (Coles, 1910).

The Remedy for Rabbitfish Stings

Members of the rabbitfish family (Siganidae) possess venomous spines (e.g., Halstead, 1978). According to Palauans the pain stemming from a rabbitfish sting is minor on the first encounter but may be quite intense on subsequent occasions. Swelling and pain begin immediately at the site of the wound (usually the hand). Within a few minutes the armpit becomes painful and red. The chemical nature of this venom has not been investigated, but information volunteered by Palauan informants suggests that it differs in different species. An individual generally experiences more pain when stung by certain species than by others. Moreover, the relative painfulness of wounds received from different species varies between victims. One person may experience particularly severe pain when stung by species A, whereas another may be relatively insensitive to species A but quite sensitive to species B.[4] (As rabbitfish are a very important food in Palau a fisherman has plenty of opportunities to get stung and thereby gradually establish his personal hierarchy of tolerances.) Some Palauans say they experience more pain when stung by certain species of rabbitfish than when stung by the more infamous stonefish.

Palauans have a remedy for rabbitfish venom which could be of pharmacological interest. The raw internal organs (or sometimes just the gall bladder) of the fish are said to alleviate the pain within three or four minutes when rubbed on the wound. Otherwise the pain is said to persist for several hours. Fishermen in the Gilbert Islands and Northern Marianas similarly employ the skin, the internal organs, or the gall bladder of rabbitfish in the same fashion. The natives of Losap Atoll, near Truk, use the gall bladder (Severance, pers. comm.). Fishermen were emphatic that the remedy works only if tissue from the same species that delivered the sting is used. (I was stung only once. As informants predicted, I felt little pain and thus had no occasion to try this remedy.)

Useless nostrums abound in the pharmacopeias of Pacific islanders just as they do in drugstores. But the details concerning this remedy make it sound plausible and thus worth further investigation. When the internal organs are rubbed on a sting, nonspecific proteolytic digestive enzymes spilled from the intestines could conceivably inactivate the toxin. But if this were so, the organs of one species of rabbitfish ought to work as well as those of any other. As the remedy is reportedly species-specific a more specific biochemical basis for its action appears likely. The fact that stings cause pain only on the second and subsequent occurrence suggest that the reaction is of an immunological nature.

Because rabbitfish mill around in dense schools while spawning and probably inadvertently sting one another occasionally, it is likely that they have evolved an immunity to the venom of their own species. If so, their tissues would contain antigens against the antitoxin in the spines, thus providing a sound biochemical basis for the prescribed treatment of their stings.

The "Tide Gauge" Eel

Helfman and Randall (1973) in their publication on Palauan fish names relate the following:

> *Tukidolch* is a congridlike eel and 10 cm in diameter which supposedly attains a length of 5.5 m. This eel lives in a burrow in mud

---

4. Banner (1977) notes conflicting reports on the degree of pain associated with rabbitfish stings. The information volunteered by Palauans suggests that these differing accounts are not really contradictory.

bottoms at depths of about four to six meters near the mangrove regions of western Babeldaob. It is said to extend itself from its burrow only at night (and hence has been seen only on moonlight nights) with its head just beneath the surface of the water. Its name means "tide gauge," in reference to its habit of maintaining its head just beneath the surface regardless of tidal oscillations.

The information I obtained on *tukidolch* differs only in that the animal was said to be somewhat smaller, found in somewhat shallower water, and its distribution restricted to small areas in the southwestern portion of Babeldaob.

The Arboreal Octopus

I have saved the strangest story for last. Many Palauans say they have seen octopi climbing trees in the mangroves. Many of the old men I talked to throughout the islands said that they had witnessed this event several times during their lives. Most of them said that the octopus climbed the *urur* tree, *Sonneratia alba*, up to a point several feet above water level, where a ring of ferns typically grows out of the trunk. Here, to make the story even more bizarre, the octopus is said to give birth to several dozen babies, which soon crawl down the tree and into the water, being preyed upon by birds and crabs as they do so.

Now if this story strikes the general reader as strange, it strikes the biologist as virtually incredible. For one thing no octopus is known to release its young directly. All octopus species whose reproductive biology is known lay eggs. In most cases the eggs hatch into pelagic larvae that live in the plankton for some days before settling to the bottom and assuming the adult form and habits. (In one species, *Octopus maya*, of the western Caribbean area, the eggs hatch directly into miniature bottom-dwelling replicas of the adults without any intermediate larval stage.) But even if Palau's octopus brooded its eggs internally and released benthic juveniles, of what conceivable advantage would a habit be that exposes both parents and offspring to heavy predation in an alien environment where they could move about and defend themselves only with great difficulty?

I have heard and read plausible accounts in various parts of the Pacific of octopi leaving the water briefly to move a short distance up the beach in pursuit of prey, such as crabs. The ancient Greeks also claimed that octopi sometimes ventured onto dry land. And Oppian maintained that octopi sometimes climbed olive trees because "they long for olives." The latter testimony is, of course, less than persuasive. While I was in Palau an octopi climbed up some concrete steps in a sea wall to a height of several feet above the water level at the Micronesian Mariculture Demonstration Center according to technicians, but it was killed before I could be summoned to witness the event. But nowhere outside of Palau, including Hawaii, Yap, Ponape, Truk, Tarawa, Saipan, and western Malaysia, where I also interviewed fishermen, could I locate so much as a rumor that octopus have babies in trees.[5]

There are three possible explanations for this story: (1) it is true; (2) it is false, but Palauan fishermen (who are quite familiar with octopi) have periodically seen something that *seemed* to them to be an octopus having babies in a tree; and (3) I was deliberately and independently misled by a considerable number of fishermen, including Ngiraklang. I am unable to favor any of these possibilities; they all strike me as improbable.

5. There is a story in Ponape that octopus sometimes enter the taro fields and climb into taro plants, but most Ponapeans I talked to treated this story simply as a legend. Palauan informants were quick to differentiate between actual observations and legends, but all of them were quite serious about the reality of the octopus story.

EPILOGUE

A culture is defined in part by the specialized knowledge it possesses. The extent to which this knowledge is retained is one measure of the strength of that culture. Today in Oceania knowledge about fishing, as well as farming, hunting, medicine, and navigation, is disappearing because younger members of island cultures are often no longer interested in mastering it. They judge it no longer useful. Why learn to fish well when nine-to-five jobs in air-conditioned offices beckon in the district center and there is an endless supply of fish in cans? Why learn to build a canoe and sail it when fiberglass boats and outboard motors can be bought and operated with little detailed preparation or knowledge? This disdain is reinforced by well-meaning educators, for the exclusion of traditional skills and knowledge from westernized school curricula in many developing countries amounts to a constant, tacit assertion that such things are not worth learning.

I have pointed out in this book and elsewhere (Johannes, 1978a, 1980) that traditional knowledge can be invaluable to Western scientists as an aid in conserving natural resources. Such an argument is not liable to motivate many young islanders to acquire it.

But there is another, more compelling reason for mastering it. Pacific island economies were traditionally self-sufficient. Detailed knowledge of the local environment and of ways to exploit it wisely was essential to this sufficiency. Today there is a heavy and growing reliance on imported technology, energy, and food. A rising influx of foreign aid, investment capital, and tourists in the past decade has stimulated much recent planning and research focusing on the further expansion of island market economies. These efforts have been based on the assumption that the growth of the world economy, on which local market economies depend, will continue. The ominous international economic climate in which these words are being written makes this assumption seem dangerous indeed.

Pacific islanders are at the end of a long and expensive supply line that now delivers much of what they once obtained within their own islands. Their economic well-being is now at the mercy of alien decision makers and impersonal market forces centered thousands of miles away in foreign capitals and trade centers. When the world economy falters, tourism, foreign aid, and foreign investment are among the first things to be affected. Pacific islanders will thus be among the first to feel the impact of a global depression. And ultimately, their market economies will be among the hardest hit. If this occurs arguments between those who espoused rapid economic progress for the Pacific islands and those who wanted to preserve traditional island cultures will seem academic indeed.

The success of an involuntary return to greater self-sufficiency would hinge largely on the extent to which traditional knowledge—knowledge gained specifically to foster self-sufficiency—has been retained. A book such as this cannot preserve such knowledge in sufficient practical detail to serve the purpose. It must be retained within the culture as a matter of personal experience.

I am not arguing that islanders should suddenly, voluntarily reduce their participation in international market economies. But if they wish the option of doing so relatively painlessly, should the time come when they have no choice in the matter, then they must strive to preserve the traditional knowledge and skills that have served well for so many centuries.

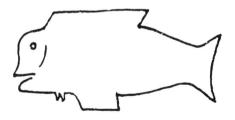

APPENDIX A

REPRODUCTIVE RHYTHMS, SPAWNING LOCATIONS, GOOD FISHING DAYS, AND SEASONAL MIGRATIONS OF REEF AND LAGOON FISHES OF PALAU AND OTHER PACIFIC ISLANDS

A number of generalizations, in addition to those discussed in the text, emerge from an examination of the information on reproductive behavior described in the following pages. They are listed here:

1. Far more reef and lagoon fishes exhibit lunar reproductive periodicity than is known to be the case for fishes of any other habitat.
2. The lunar reproductive rhythm of a species is not invariable from place to place. For example, some species that spawn around full moon at some locations spawn around new moon at others.
3. A variety of species choose precisely the same spawning locations year after year.
4. A circular milling behavior is characteristic of spawning aggregations of reef and lagoon fish of widely differing taxa, including certain mugilids (Helfrich and Allen, 1975), carangids, lethrinids, siganids, and mackerels. In some cases this motion may be set up

as a consequence of flank-nibbling of females by courting males (e.g., Hasse et al., 1977) but in general the function of this behavior is not obvious.

Except in instances where spawning grounds are already well known in Palau I have intentionally omitted information on their specific locations so as not to contribute to their overexploitation.

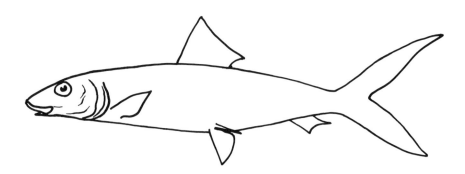

ALBULIDAE (BONEFISHES)
Albula vulpes (Linnaeus)
Bonefish
Suld

Bonefish in Palau spawn on or shortly after the new moon over sand flats near the reef edge, according to fishermen. In certain localities they also spawn around the full moon, according to one informant. The main spawning months are from January through June. Fishermen I interviewed in Tarawa, Gilbert Islands, where bonefish are the mainstay of the reef and lagoon fishery, state that bonefish form large schools in the lagoon, then migrate through island channels and form spawning aggregations over the outer reef slope from three days before until three days after the new moon throughout the year (see also Catala [1957]). At Fanning Island, Line Islands, bonefish aggregate to spawn beginning two days before the full moon, according to Gilbertese fishermen who live there (Crear, pers. comm.). These fishermen also stated that bonefish are often observed rubbing their bodies against coral rock while spawning, apparently to aid in extruding their gametes. The best fishing time (with nets) for this species is during spawning periods.

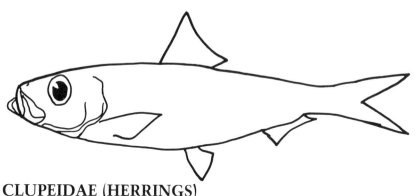

CLUPEIDAE (HERRINGS)
Herklotsichthyes punctata (Rüppell)
Herring
Mekebud

During the day *mekebud* hover in dense schools in shallow water along mangrove beaches or over sandy bottoms in the rock islands, moving only when harassed by predators. At dusk, according to fishermen, they move into deeper water to feed on zooplankton. Until 1972 *mekebud* spawned in the tidal creek at Ngeremlengui from November to April every year for as long ago as anyone in the village can remember. Spawning occurred on a high outgoing tide during midmorning. Normally schools were said to appear at the spawning site a day or two before the full moon and spawn each morning for several days starting on or about the full moon. But *mekebud* populations have been severely depleted by overfishing (see p. 83) and no spawning schools had been sighted in the creek for two years prior to my arrival in 1974.

A small school (several thousand individuals) appeared in October of 1974, however, and reportedly spawned on three consecutive days in early November, commencing two days after the full moon. The following month they reportedly spawned only on the day before the full moon. There was some indication that disturbance due to considerable boat movement in the area may have terminated spawning activities prematurely. They continued to spawn monthly around the time of the full moon through April 1975.

Spawning occurs in the upstream portion of the boat mooring area. Here the salinity is strongly stratified on a high tide. The fish are said to dive into the lower, high-salinity layer to spawn. (Poor visibility interfered with my single attempt to witness spawning.) Afterward they reportedly move to a point where a fresh-water stream enters the estuary, where they make gulping movements briefly (Palauans say they drink fresh water after spawning) before moving downstream to the fringing reef.

Large schools of *mekebud* used to spawn near Koror in seawater of full salinity in front of the old boatyard. Heavy fishing pressure has apparently greatly reduced the numbers of *mekebud* in this area. On the eastern side of Babeldaob *mekebud* reportedly spawn in the vicinity of Youlbeluu.

Mekebud are heavily preyed on by jacks, mackerel, and snappers on the fringing reef at Ngeremlengui. Their retreat to the upper portions of the tidal creek to spawn reduces this predation.

The species has a well-known migration pattern according to Palauan fishermen. Schools reportedly move northward from the Rock Islands along both sides of Babeldaob. The fish around Ngeremlengui leave in April or May, moving further northward and arriving at the northern tip of Babeldaob in May or June. Subsequently the schools pass around the northern tip and move southward down the east coast. Meanwhile other schools move northward part way up the east coast of Babeldaob. Both northward and southward moving schools disappear somewhere between Ngiwal and Ngesar. Fishermen have no explanation for their disappearance.

Whereas *mekebud* spawn at Ngeremlengui between November and May, there is some suggestion that spawning also takes place in other months elsewhere in Palau. The southern limit of *mekebud* in Palau is Ngemelis. The best fishing time for this species (with cast nets) is during their spawning periods.

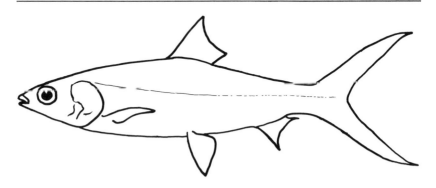

CHANIDAE (MILKFISH FAMILY)
Chanos chanos (Forskål)
Milkfish
Mesekelat, Chaol

Milkfish are among the most important food fish in the coastal tropics, annually producing more than 200,000 tons of human food in Taiwan, the Philippines, and Indonesia. Their culture in coastal ponds goes back at least 500 years in Java and 300 years in Hawaii. Yet surprisingly little is known about their reproductive cycle and pond owners must rely on fry collected by hand in shallow coastal waters to stock their ponds.

There are no published eyewitness accounts of the spawning act, but there is little doubt that some Palauans have seen milkfish spawn. For example, when I asked one Peleliuan fisherman whether he had actually witnessed the spawning act, he said that he had not, but that he *had* seen the fish do something peculiar. Periodically the members of sexually mature aggregations would suddenly "twist around one another and the water got all milky." In other words he had witnessed the spawning act without recognizing it as such.

According to fishermen, schools of mature milkfish in the vicinity of the southern tip of Peleliu swim slowly back and forth for several days while in spawning aggregations, their fins sometimes breaking the surface. (The head of the Department of Marine Resources in Palau, Mr. Toshiro Paulis, also reported encountering a school of hundreds of milkfish at this location swimming slowly and tightly packed at the surface on the morning of a full moon some years ago.) During this time the fish, which are normally extremely wary, are easily approached and speared by divers. Spawning is said to occur for about three days around the new and full moons, mainly from January through May. Observations in India (Chacko, 1950) and Hawaii (Nash and Kuo, 1976) also indicate lunar reproductive cycles for milkfish.

During 1975—1976 numbers of milkfish larvae, estimated to be between ten days and two weeks old, migrated on rising spring tides into a brackish mangrove pond on northern Peleliu during all months of the year except February when the tides were too low to allow them access. There was a pronounced peak in the number of entering larvae between early April and late June. These observations, made by A. Purmalis, a Peace Corps biologist with the Micronesian Mariculture Demonstration Center, fit well with the claim of Palauan fishermen that peak spawning activity occurs during the spring months.

Spawning reportedly occurs sporadically during the day, near the surface, in 4—10 meters of water, over the fringing reef, 100—200 meters from shore, near the edge of a steep dropoff. These observations agree well with those of Delsman (1929) who concluded, on the basis of eggs collected in plankton tows, that milkfish in an Indonesian area spawned during the day close to the coast in waters twenty to forty meters deep. Other authors suspect, however, that spawning in certain areas occurs some miles offshore (e.g., Tampi,

1957; Tjiptaminoto, 1956). This suggestion does not necessarily conflict with the observations of Delsman and Palauan fishermen. It is possible that species that spawn over the outer reef slope in fringing reef areas bounded by deep oceanic waters may move much further offshore to spawn in areas with wide, shallow continental shelves. In either case the adaptive strategy involved would appear to be the positioning of eggs and larvae over water deep enough so as to minimize predation by demersal and benthic predators.

I attempted to observe spawning aggregations at Peleliu during May and June of 1976 and March and April of 1977 without success. These failures are perplexing in view of the eyewitness reports of many different fishermen as well as the head of the Palau Department of Marine Resources concerning milkfish spawning aggregations in this area. This was the only instance in which attempts to locate a spawning aggregation of a species when and where Palauan fishermen predicted it were unsuccessful.

Adult milkfish feed on sand flats north of Peleliu and north of Babeldaob. Numerous large schools used to be seen from Peleliu all the way to Koror, but dynamiting has reportedly reduced their numbers greatly in the past ten years. Palauans state that milkfish migrate around Palau seasonally in a clockwise direction. Bagnis et al. (1972) state similarly that milkfish in Tahiti make regular seasonal migrations with which fishermen are familiar.

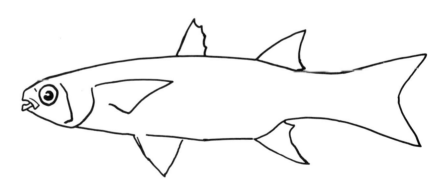

MUGILIDAE (MULLETS)

Two species of mullet exhibit lunar spawning periodicity in Palau according to fishermen. Bromhall (1954) theorized similarly that the mullet, *Mugil cephalus*, spawns near Hong Kong "near or shortly after the full moon." Malinowski (1918) described the arrival of "large shoals" of spawning mullet in the Trobriand Islands "always at full moon." Beaglehole and Beaglehole (1938) state that five unidentified species of mullet at Pukapuka were all particularly numerous around the full moon; this may also refer to spawning aggregations. Fishermen in Tarawa, Gilbert Islands, state the *Mugil macrolepis* aggregate to spawn over the ocean reef around full moon (Cross, 1978).

According to fishermen, members of two of the three common Palauan mullet species make long seasonal migrations. In other parts of the world the mullet, *Mugil cephalus*, is known to undertake extensive spawning migrations (e.g., Thompson, 1955; Titcomb, 1972).

Crenimugil crenilabis (Forskål)
Warty-lipped mullet
Kelat

This species is well known in Palau, and almost all fishermen I interviewed had seen spawning aggregations. A major spawning peak occurs around full moon, although the first major spawning of the season, usually in November, often occurs around the tenth day of the lunar month. (Similarly May et al. [1979] report that in *Polydactylus sexfilis* the first spawning of the season does not occur during the otherwise customary lunar period.) In some parts of Palau spawning reportedly occurs around the new moon.

Near Peleliu ripe individuals gather in schools over sand flats in the lagoon, then migrate out to the edge of the western fringing reef using pathways well known to fishermen, who intercept them with nets. This spawning migration usually begins three days before the full moon. Spawning takes place at the surface over the outer reef slope and begins in the evening. Similar spawning migrations occur around the full moon in Saipan, Ponape, and Truk according to fishermen. A possible explanation for why this species spawns inside the lagoon at Enewetak (Helfrich and Allen, 1975) but outside the lagoon in Palau is given by Johannes (1978b).

Seasonal clockwise migrations of schools of this species are also widely known in Palau. Schools of *kelat* move from the reef flats north of Peleliu northward up the west coast of Babeldaob. The first schools arrive at Ngeremlengui usually in November and are subsequently seen by fishermen progressively further up the coast. Schools continue to move past Ngeremlengui travelling north for about six months. When the fish reach the northern tip of Babeldaob they turn and move southward down the east coast. Today the schools are said to be much smaller and fewer than they were ten years ago because of net and dynamite fishing. The best fishing for this species (with nets) is during the spawning migration.

Chelon vaigiensis (Quoy and Gaimard)
Diamond-scale mullet
Chesau, Uluu

Spawning aggregations are reported to occur in Palau over sand in shallow water near the outer reef edge water on the fifteenth (full moon) and sixteenth days of the lunar month. Fishermen in Yap and Tarawa provided similar information. In Palau this species reportedly migrates in small schools from the Rock Islands northward along the west coast of Babeldaob, around the tip, and southward down the east coast. Members of spawning aggregations are easy to approach and net or spear. Some spawning occurs during most or all months, with a peak in March or April.

The best fishing for this species (with nets) is during the spawning migration.

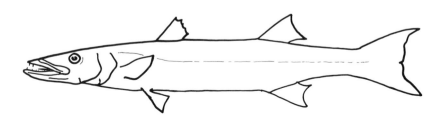

SPHYRAENIDAE (BARRACUDAS)
Sphyraena barracuda (Walbaum)
Great barracuda
Chai, Mersaod, Mordubch

This species is usually solitary except when spawning. It is commonly found outside the barrier reef but sometimes comes into shallow water. Spawning aggregations are found on the thirteenth, fourteenth, and fifteenth (full moon) days of the lunar month outside the reef. The best fishing for this species (with droplines) in Palau is during the spawning period. Gilbertese fishermen also say that the best catches of this species are made around the full moon.

Sphyraena genie (Klungzinger)
Dark-finned barracuda
Meai

This species reportedly migrates around Palau outside the reef in large, loosely associated aggregations. At Ngeremlengui the first of these aggregations appears in December, with peak numbers occurring in January. The same influx occurs progressively later as one proceeds northward up the west coast of Babeldaob, around the northern tip, and south down the east coast. What is believed to be the same wave of fish found off Ngeremlengui in December reaches Peleliu in southern Palau about seven months later. Yapese fishermen also described a clockwise migration of this species in their islands. Spawning aggregations are found on the thirteenth, fourteenth, and fifteenth (full moon) days of the lunar month outside the reef.

EXOCOETIDAE (FLYING FISHES)

In Ponape flying fish were caught in greatest numbers around moonrise and moonset for several days during and after new and full moons ("An Investigation of the Waters Adjacent to Ponape," 1937). Similarly Gilbertese fishermen told Turbott (1950) [who expresses some skepticism concerning their reports] that the best time for fishing for flying fish was either during the first three days of the new moon or at full moon and the ensuing three days. At these times the fish were said to approach the land in well-formed shoals and spawn in the breakers.

Cypselurus simus
Flying fish

A well-known spawning ground of this species is found near the town dock at Fare, Huahine, in the Society Islands (Bagnis et al., 1972). Dr. Yosihiko Sinoto has observed their spawning and provided the following description. During certain months, the most important being August, September, and October, schools of flying fish come boiling through two reef passes near the town dock at sunset, pursued by predators. The fish swim into a few inches of water close to shore, wriggle tail first into the sand and deposit their spawn much like California grunion. Spawning takes only a few seconds, after which the fish return to deeper water. No lunar periodicity was noted.

BELONIDAE (NEEDLEFISHES)
Thalassosteus appendiculatus (Klungzinger)
Keel-jawed needlefish
Sekos

Palauans and Yapese fishermen both state that this species moves into brackish mangrove waters in certain restricted areas to spawn. During the spawning period, which reportedly lasts for about a month (with no indication of lunar periodicity), the fish travel in pairs with the male swimming above the female. Similarly Breder (1959) observed pair-spawning of *Strongylura notata* around mangrove roots.

Rhynchorhamphus goergi (Valenciennes)
Long-billed needlefish

Fishermen in Tarawa, Gilbert Islands, say that the long-billed needlefish (*anaa*) migrates from the lagoon to the outer fringing reef and aggregates along the shore to spawn for three days around the new moon.

HOLOCENTRIDAE (SQUIRRELFISHES)
Myripristis spp.
Squirrelfish
Bsukl

According to Palauan fishermen the condition of the gonads of *bsukl* caught at various times of the month indicates that most spawning occurs during the first five days of the lunar month.

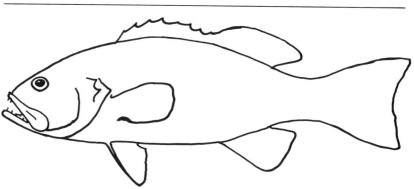

SERRANIDAE (GROUPERS)

Although groupers are essentially solitary fish, at least two western Atlantic species aggregate in specific, predictable localities to spawn (Craig, 1966; Smith, 1972; Thompson and Munro, 1974). A number of Pacific species behave similarly. Fishermen from Palau, Ponape, Truk, and Nukuoro all volunteered that a number of species of groupers aggregate to spawn mainly from March through July, typically in or near channel mouths at the outer reef slope, or on the slope itself. Similarly Severance (1976) describes groupers as spawning in a specific area on the outer reef slope at Losap Atoll in March and April. Marshallese fishermen at Majuro Atoll say that aggregations of several species of groupers occur in and near reef passages for varying periods between November and February.

Fishermen at the above-mentioned localities noted that many groupers bite avidly on the spawning grounds although their stomachs are generally empty. (Excellent fishing also occurs on the spawning grounds in the Bahamas [Smith, 1972] and off Belize [Craig, 1966], but Randall and Brock [1960] state that line fishermen in Tahiti were unable to induce spawning groupers to bite.)

Ngerumukaol Channel is the best known grouper spawning ground in Palau. It is an unusual channel in that it is a cul de sac penetrating from the outer reef edge only part way through the fringing reef. It has recently been declared off limits to fishermen during grouper spawning months.

Typically groupers that have been observed in spawning aggregations in both Pacific and Atlantic Oceans stay in the spawning area for several days or weeks before spawning. Spawning aggregations of some species in Palau reportedly form up around new moon but do not actually spawn until about full moon. Other species do just the opposite, beginning to gather in the spawning area around the full moon but not spawning until the new moon. Unfortunately I was able to observe spawning aggregations of only two species. Because grouper species are easily confused I was not able to decide in many cases what species fishermen were talking about when they described other species with lunar spawning rhythms.

Lunar spawning periodicity for the Atlantic grouper, *Epinephelus striatus*, has been noted by Smith (1972; spawning aggregation during the last quarter in the Bahamas) and by J. Gibson (pers. comm.; spawning around full moon) off Belize. In Bermuda fishermen say that *Epinephelus guttatus* spawning peaks occur on new or full moons (Burnette-Herkes, 1975). Erdman (1976) reports that *E. guttatus* spawns one or two days after the full moon off La Parquera, Puerto Rico. Randall and Brock (1960) reported that the Indo-Pacific grouper *Epinephelus merra* has a spawning peak just before and during full moon. In the Mediterranean *Epinephelus guaza* reportedly spawns around the full moon (Neill, 1966, cited in Goeden, 1978).

Palauans who reported witnessing the spawning of various groupers invariably described them as spawning in pairs rather than groups within their aggregations.

The best fishing time for a number of species of groupers (dropline, speargun) is during their spawning aggregations.

Plectropomus leopardus (Lacépède)
Leopard grouper
Mokas

Mokas reportedly begin to congregate in Ngermukaol Channel in May and June around the full moon, about two weeks before they spawn, according to fishermen. Up until a day or two before spawning commences these groupers will take baited hooks. For two or three days, starting nine or ten days after the full moon, fishermen have found that when one *mokas* is hooked another will often follow it to the surface. Consequently a second fisherman stands ready with a spear as hooked fish are pulled in. It is only at this time—just prior to the new moon—that such behavior is noted. Presumably groupers that follow their hooked brethren to the surface mistake their upward trajectory for courtship or spawning behavior. In many reef fish the spawning act involves an upward swimming movement (e.g., Johannes, 1978b). By the new moon *mokas* cease to bite. Goeden (1978) similarly found that food intake of and fishing success for this species on the Great Barrier Reef decreased markedly prior to spawning. (Here spawning also occurred near the time of the new moon in the single instance in which it was documented.)

As predicted by fishermen I saw several hundred *mokas* and *remochel* (see below) in a spawning aggregation in the mouth of Ngerumekaol Channel two days before the new moon in June 1976. I witnessed no actual spawning activity. When I returned a day after the new moon, the grouper population had declined by roughly 90 percent and the remaining fish were much warier and less approachable than they had been three days before. *Plectropomus leopardus* aggregates in channels in January and February at Majuro in the Marshall Islands, according to fishermen.

Epinephelus fuscoguttatus (Forskål)
Grouper
Remochel

In Palau spawning aggregations form around new moon and break up around full moon in May and June (for additional information see preceding paragraph). Fishermen state that the fish spawn sporadically during the entire two-week period. Similar aggregations of this species reportedly occur in November and December at Majuro, Marshall Islands.

Epinephelus merra (Bloch)
Honeycomb grouper
Mirorch

Spawning reportedly occurs around full moon inside certain channel mouths in Palau and Ponape. Randall and Brock (1960) similarly report a spawning peak just prior to the full moon for this species in Tahiti.

CARANGIDAE (JACKS)

Little has been published about the reproductive habits of carangids although the group contains a number of important food fish. In addition to the information discussed below, two other observations suggest the existence of lunar spawning rhythms in carangids. Beaglehole and Beaglehole (1938) state that the best catches of *Caranx ascencionis* and *C. sexfasciatus* are made around new and full moons at Puka Puka. Thompson and Munro (1974) report that the largest catches of a Caribbean carangid, *Caranx latus*, are made a few hours after sunset on a full moon. The greater vulnerability of many fishes to fishermen while in their spawning aggregations has been described above.

I routinely saw several species of jacks, including *Caranx mate (klspeached)* and *Carangoides fulvoguttatus (iab)*, traveling north along the inner edge of the barrier reef near Ngeremlengui on incoming tides. I witnessed these movements daily for several days prior to the full moon, but never on succeeding days, from December through May. I made no observations in other months. Most of the individuals I speared contained ripe gonads, suggesting that spawning occurred near the full moon. When groups of *C. fulvoguttatus* reached the deep channel through the barrier reef they turned and swam through it, suggesting that spawning may take place outside the reef. The best fishing for carangids in Palau (nets, spearguns) is around full moon.

In the Marshall Islands spawning aggregations of *Caranx ferdau* are said by fisherman to occur on the reef flat during winter and spring months.

Carangoides fulvoguttatus (Forskål)
Gold-spotted jack
Iab

My observations (see above) indicate that near Ngleremlengui this species spawns around full moon, probably outside the reef. In Ponape fishermen told me that this species moves from the lagoon to the outer reef slope to spawn around the new moon for several days from March to June.

Caranx melampygus (Cuvier and Valenciennes)
Bluefin jack
Oruidl

This is a wide-ranging species, found in shallow water on barrier and fringing reefs and in deeper water in the lagoon and outside the barrier reef. It moves up onto the reef flats at night to feed. On the day before the new moon of April 1977 I saw about ten schools of this species, each containing several dozen fish, moving south down the east and west sides of Peleliu. The following morning off the southern tip of Peleliu I saw an aggregation consisting of more than one thousand individuals. Judging by its size and location this aggregation may have included the smaller schools I had seen the day before. The fish I speared from the school contained ripe gonads.

This appeared to be a spawning aggregation. Its existence at this time and location had been predicted by Peleliuan fishermen. I saw no spawning.

On the same day there was also a school of unidentified jacks in the same area, the members of which swam in pairs, one above the other, the uppermost individual much darker than its companion, apparently displaying courtship behavior. (Talbot and Williams [1956] report similar sexual dichromatism in *Caranx ignobilis*, and Gooding and Magnuson [1967] observed similar behavior and color differences in *Coryphaena hippurus*.) The aggregations of both species were present throughout the day at this location. The next morning—the day after the new moon—at dawn, all but a few members of both species had left. Possibly they had spawned the night before.

Caranx ignobilis (Forskål)
Jack
Cherobk, Chederobk

Von Westernhagen (1974) describes this species as spawning in pairs, within large aggregations, during the day, over a sand bottom in twenty to forty meters of water, around slack tide, on patch reefs, in Indonesian waters. He noted that they were easy to approach while in these aggregations. Palauan fishermen in contrast, are emphatic that spawning aggregations of this species, readily recognizable because of its bulging head and large size, occur in *shallow* water on outer fringing and barrier reef flats. (Tahitian fishermen [Bagnis et al., 1972] and Palauans note that these fish also have the unusual habit of running toward shallow rather than deep water when speared, sometimes stranding themselves.)

In both Tahiti (Bagnis et al., 1972) and Palau spawning aggregations are seen around both new and full moons. Williams (1965) reported shoals of *C. ignobilis* off East Africa running ripe at the full moon.

Caranx mate (Cuvier and Valenciennes)
Yellowtail scad
Klspeached

Spawns in small schools on full moon from November through April and perhaps other months. This testimony by fishermen is supported by my observations (see *Carangidae* above).

Gnathanodon speciosus (Forskål)
Golden jack
Wii

At Ngeremlengui this species lives mainly in the lagoon and travels in small schools. It forms large spawning aggregations according to fishermen and moves into shallow water on both the fringing reef flat and the inner barrier reef flat during daylight hours where it falls prey to *kesokes* nets and thrown spears. These aggregations occur around full moon from November to May and perhaps other months. Some spawning may also occur around the new moon, according to fishermen.

Elegatus bipinnulatus (Quoy and Gaimard)
Rainbow runner
Desui

As an adult this species is pelagic and is usually found over deep water. When rainbow runners are seen by divers they are often swimming individually or in small groups and are usually too wary and move too fast to be easily speared. However, on the day before the full moon in June 1976 Daryl Gray, a Peace Corps biologist, and I encountered a school of about 200 adults in about 5 meters of water over the fringing reef along the east coast of Peleliu. These fish were swimming slowly in a tight clockwise circle, the fish distributed through the upper half of the water column.

We were surprised to find that we could approach this school with ease and we speared two individuals. One was a male, running ripe. The other was a ripe female. The stomachs of both were full of squid beaks, isopods, and small unidentified fishes. We believe that this was a spawning aggregation. No Palauan fishermen we subsequently interviewed had ever witnessed this behavior in rainbow runners, however. Only one fisherman volunteered any information on spawning for this species; Ngiraklang said that, based on the condition of the gonads, he believed this fish spawned around the full moon. This is consistent with our observations above.

Scomberoides sancti-petri (Jordan and Evermann)
Leatherback

In Tarawa Atoll, according to Gilbertese fishermen, this species (*nari*) migrates from the lagoon to the outer reef slope to spawn 5–7 days after the full moon.

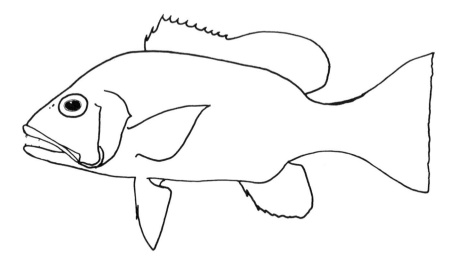

LUTJANIDAE (SNAPPERS)

The three species of the genus *Lutjanus* whose spawning times are known to Palauans all spawn on or shortly before the full moon according to fishermen. *Lutjanus griseus* in the Florida Keys is also thought to spawn around full moon (Starck and Schroeder, 1970), as does *Lutjanus synagris* in Cuba (Reshetnikov and Claro, 1976), and *Lutjanus vaigiensis* in Tahiti (Randall and Brock, 1960). In the Gilbert Islands, however, aggregations of ripe *L. vaigiensis* occur starting about two days before the new moon according to fishermen (Crear, pers. comm.).

Lutjanus argentimaculatus (Forskål)
Mangrove snapper
Kedesauliengel, Kedesau

L. *argentimaculatus* is found in many different habitats but is particularly abundant in mangrove channels. It moves into deep water in the lagoon and over the outer reef slope to spawn from the tenth to the fifteenth (full moon) throughout most or all of the year. Fishermen in Yap also described this species as a full moon spawner. Peak spawning occurs in late spring and early summer. Line and net fishing is particularly good for this species and for *Lutjanus bohar* (see below) on the eleventh, twelfth, and thirteenth nights of the lunar months while they are spawning, and these nights are sometimes referred to in Palau as "Kedesau nights." Fish caught during these nights usually have empty stomachs.

Lutjanus bohar (Forskål)
Red snapper
Kedesau

Red snappers live along the outer reef slope and in deep channels and passes. Spawners aggregate on the outer reef slope between the tenth and fifteenth (full moon) day of the lunar month. Peak spawning activity is from April through July, but there is probably some spawning in all months according to fishermen. Excellent line fishing for this species occurs during spawning nights. This species seems to move around more than most other snappers (with the exception of *Aprion virescens*) on the reef slope, traveling in small schools.

Lutjanus gibbus (Forskål)
Snapper
Keremlal

Keremlal live in the lagoon and on the outer reef slope. Spawning aggregations in Palau occur "in blue water close to the outer reef edge" from the tenth to the fifteenth (full moon) days of the lunar month throughout the year. The best fishing (dropline) for this species is during and shortly after the full moon according to Palauan and Gilbertese fishermen.

Aprion virescens (Cuvier and Valenciennes)
Jobfish
Chudel

Jobfish are found along outer and inner reef slopes, coming into shallow water to feed. They travel in groups of one to five individuals. "If five travel together one of them is often much smaller than the rest." They spawn along the outer reef slope during the first five days of the lunar month from January through May and perhaps other months. Similar spawning aggregations in New Caledonia are mentioned by Fourmanoir and Labout (1976).

This species moves around Babeldaob in clockwise direction, favoring the leeward side of the island (i.e., the west side from November to April, the east side from May to October).

Symphorus spilurus (Günther)
Blue-lined sea bream
Chedui

During the spawning months of April through July *chedui* aggregate in large schools. In May of 1976 and April of 1977 I saw such a school at the best known aggregation site in Palau in forty feet of water at the edge of the reef dropoff off the eastern shore of Peleliu. More than 1,000 fish, each weighing several pounds, were in a densely packed, almost motionless aggregation. The fish could easily be approached to within a few feet before the nearest fish made any effort to move aside.

Peleliuan fishermen state that during the months of March, April, and May *chedui* move out from inside the reef to this area several days before the new moon, stay there until they spawn around the full moon, and return to their individual haunts just after the full moon. During the early morning the aggregation is found in ten to fifteen feet of water near shore, moving out to deeper water later in the morning. Similar spawning aggregations of the related species, *Symphorus nematophorus*, in New Caledonia are mentioned by Fourmanoir and Labout (1976).

Such schools are easily decimated by spearfishing. Once such a school has been fished out in an area, it will not recover in subsequent years according to Ngiraklang. The population density of this fairly large and conspicuously colored fish seems very low on the reefs of Palau; one spawning aggregation must draw its members from a radius, I would guess, of at least several miles. This, plus Ngiraklang's observations, suggest that the species may be unusually vulnerable to local extinction.

POMADASYIDAE (SWEETLIPS)
Plectorhynchus obscurus (Günther)
Sweetlips
Bikl

According to Palauans this species spawns only once a year in April–May when spawning aggregations are seen around the new moon on the outer reef slope.

Plectorhynchus goldmani (Bleeker)
Diagonal-banded sweetlips
Yaos

Spawning aggregations reported around new moon.

LETHRINIDAE (EMPERORS)
Lethrinus ramak (Forskål)/(identified by Torao Sato)
Emperor
Chudch

During the day *chudch* are found singly on the fringing reef at Ngeremlengui. They spawn at the outer. edge of the fringing reef sometime after dark on the first five days of the lunar month from November through April and perhaps other months. *Chudch* are said to feed only at night.

The best catches of this species (nets, droplines) are made during the spawning period.

Lethrinus harak (Bleeker) (identified by Torao Sato)
Emperor
Itotch

These fish travel singly or in small schools in the mangroves and along the fringing reef at Ngeremlengui, feeding only during the day. They aggregate to spawn during the first five days of the lunar month in most or all months. At dusk the aggregations move out into the lagoon while swimming in a circle near the surface, according to fishermen, and spawn sometime during the night.

They seldom bite during their spawning days, but bite particularly avidly for three days immediately afterward.

Lethrinus lentjan (Smith) (identified by Torao Sato)
Emperor
Metngui

Except when in spawning aggregations this species is found singly in deeper lagoon waters and outside the reef. Spawning aggregations are seen only during the first half of the lunar month, but there is disagreement among fishermen as to when actual spawning takes place.

Lethrinus sp.
Emperor
Eluikl

This is a nocturnal feeder which lives on the barrier reef at Ngeremlengui where it travels singly. Spawning aggregations are seen by fishermen at the outer edge of the fringing reef and spawning occurs during the first five nights of the lunar month throughout much of the year with a peak between April and June.

The best fishing for this species with nets occurs during the spawning period. With hook and line fishing it is best around full moon at night "even when the sky is overcast."

Lethrinus miniatus (Forster)
Porgy
Mlangmud

This species travels in large schools. The largest numbers are found on the barrier reef flat. Catches had greatly diminished at Ngeremlengui in the early 1970s, but in 1975 *mlangmud* was nonetheless still one of the five most important food fish there. *Mlangmud* reportedly spawns through the year from new moon through the fifth day of the lunar month along the outer and inner edges of the barrier reef. They are caught on droplines both day and night.

The best dropline catches are made around dusk on spawning nights after they have spawned. Bagnis et al. (1972) note similarly that in Tahiti these fish bite voraciously for four or five days after the full moon at which time they are in large schools.

Lethrinus microdon (Cuvier and Valenciennes)
Porgy
Mechur

Mechur travel on the barrier reef and along the reef slopes in small schools in which *mlangmud* predominate. They spawn near the outer and inner edge of the barrier reef during the first five nights of the lunar month throughout much of the year, with a peak in April. They feed sporadically at night as well as during the day. The best dropline catches are made around dusk on spawning nights, particularly on the night of the new moon.

Monotaxis grandoculis (Forskål)
Porgy
Besechaml

Monotaxis grandoculis are often caught in traps at night when they move into shallow water to feed. Spawning aggregations are found near the bottom of reef slopes on the new moon and for the following four days according to Ngiraklang. Line fishing is best during this time. Bagnis et al. (1972) state similarly that this species is "most abundant" in the Society Islands on the fourth and fifth nights of the lunar month.

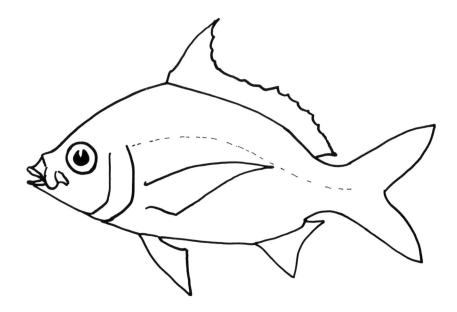

GERREIDAE (MOHARRAS)

Both species of gerreids known to Palauans have lunar spawning cycles. An unidentified species spawns in the evening around new moon in Ponape according to fishermen. The best fishing for gerreids (with nets) occurs while they are in their spawning aggregations.

Gerres abbreviatus (Bleeker)
Moharra
Chedochd

These fish migrate from mangroves and sand flats to the outer reef edge where they spawn in schools on the fifteenth (full moon) and sixteenth days of the lunar month. Small numbers of gravid *chedochd* sometimes mix in with spawning aggregations of the closely related *Kotikw, Gerres oblongus.*

Gerres oblongus (Cuvier and Valenciennes)
Moharra
Kotikw

Kotikw gather in large schools inside the reef in shallow sandy areas before migrating to specific sandy spots near the outer reef edge or channel edge. The migration out through Denges Pass to the outer reef flat southeast of Ngerong Island is particularly large and well known, although dynamiting is said to have reduced this run considerably in recent years. Spawning occurs during late afternoon and proceeds for several hours.

Spawning occurs during most months of the year on the fourteenth, fifteenth (full moon), and sixteenth days of the lunar month. I witnessed the gill netting of several hundred running ripe *kotikw* in the early evening on the full moon of April 1977 on the sand flat northwest of Ngerong Island. Fishermen are uniformly emphatic that *kotikw* will not spawn unless there has been a heavy rain and a short period of rough weather a day or two prior to their spawning period. Because it rains and blows so frequently in Palau, this not an easy assertion to investigate.

Gerres argyrus (Bloch and Schneider)
Moharra

According to Gilbertese fishermen large aggregations of *G. argyrus (ninimai)* "filled with eggs" are encountered around full moon.

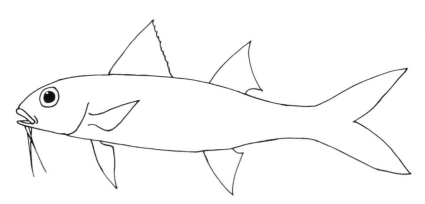

MULLIDAE (GOATFISHES)

Ponapean fishermen say that in their waters an unidentified goatfish spawns in pairs over seagrass beds at high tide at dusk around new moon. Beaglehole and Beaglehole (1938) state that the best catches of a species of goatfish ("vete") are made around the new moon at Puka Puka.

Mulloidichthys flavolineatus (= *samoensis*) (Lacépède)
Goatfish
Dech, Chemisech

This species spawns over shallow sandy areas near the reef edge for several days beginning one to several days after the new moon according to Palauans. The best fishing for this species in Palau (with nets) is during the spawning period.

Upeneus arge (Jordan and Evermann)
Goatfish

According to Gilbert Islands fishermen *Upeneus arge* (*maebo*) migrate from Tarawa lagoon to the outer reef edge to spawn around the new moon for three to six days.

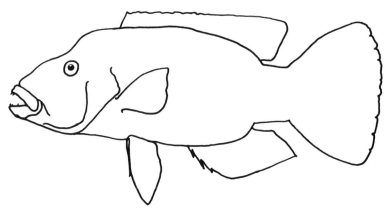

LABRIDAE (WRASSES)

Roede (1972) reported lunar reproductive periodicity in a number of Caribbean labrids (see also Randall and Randall, 1963; Feddern, 1965). Some species, however, exhibit no apparent lunar reproductive rhythm (e.g., Warner and Robertson, 1978).

Cheilinus undulatus (Rüppell)
Bumphead wrasse, Napoleon wrasse
Mamel

Palauans are uncertain as to when *mamel* spawn, and no one I talked to had ever knowingly witnessed the spawning act. Similarly, and despite widespread value of this species as food in the Indo-Pacific, its spawning has never been described in the scientific literature. However, Tuamotuan fishermen state that in March pairs of *C. undulatus* "swim 'like fools' and 'play' and rise, spiralling from the depth to the surface, then quickly dive almost vertically toward the sand" (Ottino and Plessis, 1972; translated from French). These observers were almost certainly, unwittingly, describing the spawning act; published accounts of pair-spawning in other species of reef wrasses fit this description closely (e.g., Robertson and Hoffman, 1977).

Choerodon anchorago (Bloch)
Yellow-cheeked wrasse
Budech

This species lives in rocky or coral-covered areas of the fringing reef. Spawning aggregations are seen at the outer edge of the fringing reef around new and full moon in January with smaller spawning aggregations reported in February and March.

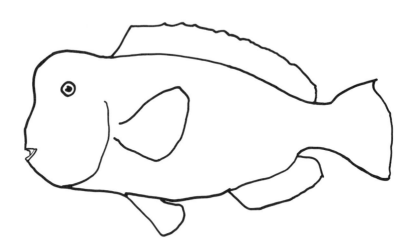

SCARIDAE (PARROTFISHES)

According to Ngiraklang a number of species of parrotfish, including *Scarus harid* and *S. gibbus*, frequently aggregate at the extremities of underwater promontories on the outer reef slope to spawn. Randall and Randall (1963) described two species of Caribbean parrotfishes that spawn at such locations. The possible advantage of such spawning locations is discussed on page 45 .

A number of larger species of parrotfishes reportedly spawn between the sixth and tenth day of the lunar month in Palau. But the taxonomy of parrotfishes is sufficiently confusing that I was only able to identify one of these species clearly (see below). The period when parrotfish spawn in Palau has a special name, *lmall*, and net fishing for them is best at this time. It is perhaps not coincidental that the natives of Raroia say that the ninth day of the lunar month is particularly good for catching parrotfish (Danielsson, 1956).

Ponapean fishermen report, however, that two species of parrotfish spawn around the new moon in their waters. During this period in March, April, and May *momei* (*Scarus harid*) and another unidentified species of parrotfish called *mal* aggregate to spawn at certain locations on the barrier reef flat.

Fishermen in the Tuamotus say that some parrotfishes make migrations within the lagoon and that the direction of these migrations reverse with the lunar phases (Ottino and Plessis, 1972). Palauans were not aware of such migrations although they mentioned that there seemed to be considerable seasonal variation in the abundance of *Bolbometopon muricatus* in certain areas. Fishermen in Ponape volunteered the same information.

Bolbometopon muricatus (Cuvier and Valenciennes)
Bumphead parrotfish
Kemedukl

Palauans have never seen this fish spawn. A number of fishermen noted, however, that *kemedukl* contain well-developed eggs from the first to the ninth of the lunar month and that they appear to spawn on the eighth and ninth days, probably after sunset. Ngiraklang said that spawning aggregations could be found near the inner entrance to the barrier reef channel near Ngeremlengui starting on the eighth day of the lunar month. I saw such aggregations, beginning on the eighth day of the lunar month, in the spot he indicated but was unable to get close enough to spear any fish to examine their gonads.

ACANTHURIDAE (SURGEONFISHES)

At least five species of acanthurids exhibit lunar spawning periodicity in Palau according to fishermen. Randall (1961*a*) notes

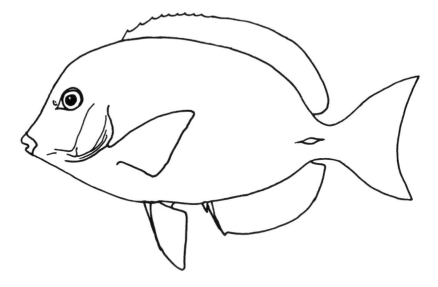

several other species that appear to exhibit lunar spawning period-icity in Tahiti. Spawning aggregations of *Naso hexacanthus* and *N. brevirostris* occur throughout most of the year during neap tides near some channel mouths in the Marshall Islands according to fishermen.

Acanthurus mata (Cuvier)
Surgeonfish
Chesengel

These fish travel in schools that join to become very large schools during the spawning period. Spawning aggregations are seen on the reef flat around both new and full moons in May and perhaps other months. *Chesengel* feed in shallow sandy reef areas during the day and move down the reef slope to rest on the bottom at night. The best fishing time for *chesengel* (with nets) is during their spawning period.

Acanthurus triostegus (Linnaeus)
Convict surgeonfish
Chelas

Spawning aggregations have been seen off the outer reef edge near the outer entrance to the channel through the barrier reef at

Ngeremlengui (Toachel m'lengui). Spawning is said to take place in the evening between the fourth and tenth of the lunar month. Randall (1961b) describes a somewhat similar lunar spawning rhythm for *Acanthurus triostegus sandvicensis* in Hawaii. Ninety percent of the running ripe individuals he collected were obtained between the fourth and the seventeenth days of the lunar month. In Palau spawning occurs between May and August and perhaps to a lesser extent in other months.

A diver in Ponape twice witnessed the spawning of this species in about ten feet of water near the outer entrance to a large channel through the barrier reef. Periodically the school would rush upward suddenly, releasing eggs and milt as the fish converged at the top of their rush. This spawning pattern has been observed in a variety of reef fish, including other surgeonfish (Randall, 1961a). Spawning occurred in the early evening. Sharks and jacks harassed the school as it spawned.

Acanthurus xanthopterus (Cuvier and Valenciennes)
Ring-tailed surgeonfish
Mesekuuk

This species appears to move around considerably; fishermen noted that there is considerable seasonal variation in abundance. Fishermen in Ponape volunteered the same information. Spawning in Palau reportedly occurs around both new and full moons from January through May.

Acanthurus lineatus (Linnaeus)
Blue-lined surgeonfish
Belai

I observed the repeated spawning of *Acanthurus lineatus* for about one-half hour (it was continuing as I left the area) starting at 6:45 A.M. on the morning of the new moon in April 1977 at the southern tip of Peleliu Island. Groups of five to fifteen fish would spiral slowly upward in the water column, rising about three meters above rock outcroppings on the outer reef flat or above the edge of the outer reef drop-off. At the top of this trajectory the fish released clouds of gametes, then returned unhurriedly to positions near the bottom. Many times the fish made false starts, aborting their upward movement. The following morning at the same time *belai* were still to be seen at these locations but I saw no spawning.

Naso unicornis (Forskål)
Long-snouted unicornfish
Chum

Chum move in schools along the outer reef slope, traveling considerable distances. Throughout Palau fishermen noted great seasonal variations in abundance of this species. Ponapean fishermen volunteered the same information, and Bagnis et al. (1972) noted considerable seasonal variation in its abundance in the Society Islands. *Chum* are abundant at Ngeremlengui beginning in December, around northern Babeldaob beginning in August, and near Peleliu beginning in March. Judging by the condition of the gonads, according to fishermen, *chum* spawn around both new and full moons.

Naso annulatus (Quoy and Gaimard)
Ring-tailed unicornfish
Mengai

Naso annulatus lives and spawns along the inner barrier reef edge and slope near Ngeremlengui, spawning on new and full moon in May and perhaps other months according to fishermen.

SIGANIDAE (RABBITFISHES)

There are at least ten species of rabbitfishes in Palau. Rabbitfish often travel in multispecies schools, particularly during the spawning period. In Guam, about 550 miles northeast of Palau, the pelagic juveniles of two species of rabbitfish, *Siganus argenteus* and *S. spinus*, swarm onto the reef flats in very large numbers around the third lunar quarter in April and May (Tsuda and Bryan, 1973). Although the same two species occur in Palau, such massive inshore migrations of juveniles are unknown there, possibly because the species is less abundant there. Kramer (1903) mentions similar schools of juvenile siganids (and acanthurids) moving from the open ocean into the lagoon in Samoa. The natives of Lau, Solomon Islands, say that the young of unidentified siganids migrate inshore about fifteen days after spawning has occurred (Akimichi, 1978).

In Palau at least four species of rabbitfish have lunar spawning rhythms. Other siganids reported to have lunar reproductive rhythms are *S. vermiculatus* (Popper et al., 1976), *S. rivulatus* and *S. luridus* (Popper et al., 1979), and *S. spinus* (Aquaculture Team of the Centre Oceanologique du Pacifique, pers. comm.). *S. doliatus* spawns around the new moon in Ponape according to fishermen. Unidentified rabbitfish undertake spawning migrations in the New Hebrides four to five days after the new moon (Hallier, 1977). Similarly, unidentified siganids are said by Solomon Island fishermen to migrate "to" (toward?) the open sea to spawn four days after the new moon (Akimichi, 1978). In Palau the best catches of rabbitfishes are made with nets during spawning periods.

Siganus canaliculatus (Park)
Rabbitfish
Meas

Some of the aspects of *meas* biology known to Palauan fishermen are described by Hasse et al. (1977). *Meas* inhabit shallow seagrass flats near mangrove areas, retreating to reef depressions and channels during low tides. They abandon these areas beginning about a week before they spawn and migrate to one of at least twenty known locations along the outer edge of fringing or barrier reefs. The departure of *meas* on their spawning migration is so abrupt that one day thousands may be caught, whereas the next day few or none are caught. According to Ngiraklang these fish travel to their spawning grounds in monosexual schools and the females generally arrive first.

According to fishermen spawning commences three to six days after the new moon, starting a day or so later each month during the peak spawning season. Most spawning activity occurs between February and June, although a few *meas* spawn in other months. These lunar and seasonal spawning rhythms are in good agreement with those described by Manacop (1937) for the Philippines and by fishermen from Yap. Hasse et al. (1977) point out that the peak spawning periods for *S. canaliculatus* in Palau coincide with the times of the year and of the lunar month when extreme low tides occur around sunset. Peak numbers of spawners generally occur on the third day of spawning according to fishermen.

Spawning is said to occur around low slack tide in the middle to late afternoon and evening. Spawning is also occasionally seen in the early morning. One female in captivity spawned repeatedly in successive months (Hasse et al., 1977). The eggs are typically laid in shallow breaking surf at the reef edge. According to Ngiraklang spawning occurs somewhat shoreward of this area if the surf is particularly rough. Prior to spawning the fish reportedly tend to mill in slow, tight circles. They form pairs just before spawning. The eggs are adhesive and denser than seawater, so they sink and may adhere to the substrate.

For the first three weeks after hatching the larvae are apparently pelagic. About twenty-one days after hatching juveniles appear in schools on the grass flats on at least thirty-eight locations known to fishermen (McVey, pers. comm.).

The large March or April spawning run near Airai used to be the object of considerable festivity and of pilgrimages from other parts of Palau, but overfishing has greatly reduced this run in recent years. Airai now has a municipal ordinance prohibiting fishing for *meas* with *kesokes* nets during the spawning season. The fish that spawn at Airai came from Aimelik and Koror areas, according to fishermen, and can be seen migrating to and from the spawning grounds through the bridge channel.

Meas are among what Palauans call the "stupid" spawners; they can be approached with unusual ease while in their spawning aggregations. This species and the closely related *klsbuul (Siganus lineatus)* were by far the most important fish by volume in Ngerem-lengui catches in the 1950s according to fishermen. But intensive fishing with small mesh nets has led to a very marked decline in numbers and mean size of *meas* in recent years. Fishermen in other parts of Palau report similar declines in landings.

Meas feed on seagrass during the day. They take almost no food in spawning aggregations but reportedly feed particularly voraciously immediately after spawning.

Siganus lineatus (Cuvier and Valenciennes)
Rabbitfish
Klsbuul

Klsbuul are abundant, easily caught, and well liked in Palau. I was therefore surprised when, over a period of ten months, I was unable to locate a Palauan fisherman who was certain where this species spawned. Then in Ngiwal I met Temol, a fisherman known for his ability to dive deep and stay down a long time. He told me that ripe *klsbuul* formed large, docile schools in the mangroves, then migrated seaward through a channel in the fringing reef off Ngiwal, aggregating near the bottom over a sandy area on the outer reef slope in the channel mouth in about seventy feet of water. These aggregations formed on the ninth and tenth days of the lunar months. When the fish returned to shallow water they were spent. Few spearfishermen frequented this area and even fewer dive deep enough so that they would have been likely to see this phenomenon. (Although Palauans are skilled divers in most ways, the majority do not dive deeper than about thirty feet.)

Darrel Gray, a Peace Corps biologist stationed in Ngiwal, regularly speared *klsbuul* on the adjacent fringing reef and kept a record of his observations subsequent to my interview with Temol. He noted that in June 1976 all the *klsbuul* he speared contained well-developed gonads until the morning of the tenth day of the lunar month, at which time all the fish he speared were spent.

Peak numbers of ripe *klsbuul* are reportedly caught between March and June and again around November, with some ripe fish found throughout the year. Individuals kept in captivity spawned several days before the full moon during three successive months (Bryan, pers. comm.). In New Caledonia migrations of gravid *S. lineatus* and another unidentified species of rabbitfish have been observed moving into the bay of Port Sandwich four or five days after the new moon, returning spent two days later (Hallier, pers. comm.).

Siganus argenteus (= rostratus) (Quoy and Gaimard)
Rabbitfish
Beduut

This rabbitfish lives over sand in areas rich in coral growth. In Palau spawning occurs several days after both new and full moons on the outer reef flat according to fishermen. Spawning occurs mainly in March, April, and May. In Tahiti *S. argenteus* spawns for several days commencing four or five days after the full moon (Aquaculture Team of the Centre Oceanologique dur Pacifique, pers. comm.). In the Red Sea this species is believed to spawn at the time of the new moon (Popper et al., 1979). At Majuro Atoll *Siganus argenteus* reportedly make spawning migrations of several miles and form spawning aggregations for several days commencing a few days after the new moon. Around Saipan this species reportedly aggregates to spawn around the time of the new moon.

Siganus punctatus (Bloch and Schneider)
Rabbitfish
Bebael

Bebael live on the outer reef flat near the reef crest. They usually travel in pairs except during spawning periods when they form large schools. They spawn around low tide near the outer reef edge just shoreward of where the waves break. Spawning occurs around the new moon and also around the full moon in some areas, according to Palauan fishermen. In Ponape spawning reportedly occurs on the new moon and for several days thereafter. Peak spawning in Palau is said to occur in October or November.

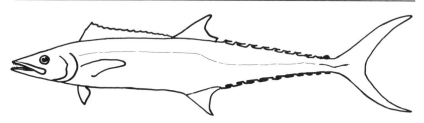

SCOMBRIDAE (MACKERELS AND TUNAS)

Acanthocybium solandri (Cuvier)
Wahoo
Keskas, Mersad

The fishermen of both Palau and Tarawa, Gilbert Islands, say this fish is caught on trolling lures in unusually large numbers around the full moon and that judging by the state of the gonads they may spawn around this time. Particularly large catches are made in Palau south of Peleliu around the full moon in April, May, and June.

Grammatorcynus bicarinatus (Quoy and Gaimard)
Shark mackerel, Double line mackerel
Biturchturch, Mokorkor

According to Silas (1963) "no information is available about prespawning or spawning" of this species. Palauans could provide me with little information on this fish, but I obtained the following information while spearfishing.

Schools of shark mackerel were frequently seen in two to five meters of water a few meters from the edge of the inner barrier reef slope immediately south of *Toachel m'lengui*, the deep channel through the barrier reef west of Ngeremlengui. Over the course of about forty dives made at this spot during a six month period I saw several hundred schools containing two to several hundred fish. All were swimming steadily north along the reef edge.

These schools appeared only on an incoming tide. The first schools of the day would often appear several minutes after low slack tide. I have no idea where the fish were during ebb tides. Such schools were not seen from the seventeenth to about the twenty-fifth day of the lunar month. From about the twenty-sixth of one lunar month to the seventeenth of the next, schools were seen each day. I obtained specimens by spearing. Most were sexually mature. During each three-week period when these fish were seen, their gonads apparently grew rapidly. Fish sampled early in these periods generally had small fish in their stomachs. Feeding apparently slowed, then ceased as their gonads matured; for the last four days (thirteenth to seventeenth) in each run their stomachs were empty or nearly empty.

During the last several days of this period the gonads occupied virtually all available space in the body cavity, causing an abdominal bulge that could easily be seen as the fish swam by. I never obtained fish that were running ripe, but the above observations suggest that spawning occurred around full moon.

These fish were often observed to turn into the deep channel through the barrier reef and swim toward the outer reef, suggesting that they may spawn outside the reef. On one occasion (in my absence) hundreds of shark mackerel were caught in a two-hour period by a tuna boat crew at the outer entrance to the deep channel, *Toachel m'lengui*.

Individuals containing roe were forty-three centimeters in fork length or larger (maximum observed fifty centimeters). Occasionally smaller individuals (thirty-three to forty centimeters fork length) were seen, usually but not always traveling in company of larger individuals. The former had small gonads. Gonadal growth, however, proceeded rapidly in this size class during each three-week observation period and it appeared as if these individuals would spawn during the subsequent lunar month. It also appeared as if these fish grew several centimeters in length per month judging by the size of "this month's spawners" compared to "next month's spawners."

Of the individuals whose sex was recorded, twenty were females and only four were males. Fish with mature gonads occurred from late December to July. I made no observations in other months.

One of the Palauan names for this species, *biturchturch*, means "urine" and refers to the ammonia smell given off by these fish if they are boiled without removing the kidney tissue from along the backbone. (The English name, shark mackerel, also refers to the shark-like ammonia smell.) The smell disappears if the fish is filetted and the backbone discarded. The flesh is mild and pleasantly flavored and the roe are of particularly good flavor. Palauans are somewhat disdainful of this fish because it is not customary to filet fish and it is a nuisance to have to scrape out the kidney carefully to get rid of the strong smell.

Scomberomorus commersoni (Lacépède)
Spanish mackerel
Ngelngal

Spanish mackerel migrate clockwise around Babeldaob according to fishermen. The first wave of fish in this annual migration appear off Ngeremlengui in December. Yapese fishermen also volunteered that this fish makes a clockwise migration in their waters. Extensive migrations of these fish also occur along continental coastlines (e.g., Copley and Allfree, 1950; Roughley, 1966). The spawning season of this species is apparently prolonged but I was unable to obtain more specific information.

Trachurus boops (Cuvier and Valenciennes)
Ox-eye scad
Terekrik

These fish are often found on the fringing reef by day. At night they move out into the lagoon to feed on zooplankton according to Ngiraklang. *Terekrik* reportedly spawn around the tenth and eleventh days of the lunar month from November through April and possibly in other months. Near Ngeremlengui schools spawn at the surface over deep water near the outer edge of the fringing reef starting in the early evening on a falling tide. According to fishermen spawning occurs while large schools swim in tight circles. Such schools are often harassed by large predators including several species of carangids.

The best catches of *terekrik* (with nets) are made during the spawning period. Despite heavy fishing pressure their populations do not seem to be diminishing according to fishermen.

BALISTIDAE (TRIGGERFISH)

Palauans say all triggerfish build nests and guard their eggs. I observed the nests and nest-guarding of one species.

Pseudobalistes flavimarginatus (Rüppell)
Triggerfish
Dukl

This species excavates nests in sand, in or near channels through the reef. These nests were particularly abundant on the inner edge of the reef flat immediately north of the lagoon end of *Toachel m'lengui*, the channel through the fringing reef near *Ngeremlengui*.

The nests are guarded by adult fish. When I approached such nests the adults always withdrew beyond spear range and small triggerfish or butterfly fish quickly took advantage of this opportunity to eat the eggs. As soon as I withdrew the adults streaked back to the nest and drove out the egg predators.

The nests were about two feet deep in the center and about eight feet wide. They are made, according to Ngiraklang, by the fish turning on its side near the bottom and wriggling rapidly (much as a salmon digs its nest). The eggs are deposited in a single, spongy, fist-sized cluster in the center of the nest. They are only slightly negatively buoyant; if some are broken off from the cluster, a moderate wave surge will sweep them out of the nest. The eggs are weighted down by several small chunks of dead coral piled on top of the cluster. Although I did not witness adults carrying rubble to the nests in their mouths for this purpose, there can be little doubt that this is what occurs; such fragments are often absent from the vicinity of the nests except on top of the eggs. This is an unusual example of a fish using a tool—as ballast. (Fricke [1971] describes how triggerfish unpile rocks from around sea urchins in order to eat them.)

I removed the eggs from one nest and measured them. The eggs were roughly spherical, transparent, and without obvious visible inclusions. The mean diameter was 0.55 mm, and there were roughly 430,000 in the cluster.

This is close to the lower size limit for fish eggs (Breder and Rosen, 1966). Williams (1959) showed that fish that provide parental care for their eggs do not necessarily produce only small numbers of large eggs, contrary to the belief of earlier workers. The present observations support these findings.

I found nests with eggs only during the periods from three days before to one day after new and full moons. Apparently egg-laying precedes new and full moons and hatching takes place at approximately the time of maximum tidal flushing. I found eggs in the months of November, December, March, April, and May. I did not look for nests during other months.

Predators that eat triggerfish eggs are apparently active only during the day. *Dukl* rest in holes at night, leaving their nests unattended. Whereas egg predation was common when *dukl* were driven from their nests during the day, I never witnessed it at night.

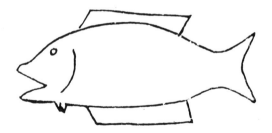

APPENDIX B

CRUSTACEANS: LUNAR REPRODUCTIVE RHYTHMS AND MISCELLANEOUS BIOLOGICAL OBSERVATIONS

The lunar larval-release rhythm of the land crab, *Cardisoma hirtipes* (*rekung el beab*), was described in Chapter 3. This crab makes burrows ranging in length from one to six feet and varying in depth depending partly on the height of the water table. A burrowing crab that hits water creates a bend in the burrow so that it does not extend into the water. During long heavy rains crabs are forced out of their burrows by flooding and this is a good time for Palauans to catch them.

No one has apparently seen these crabs mate. Molted shells, however, are frequently found in hollows beneath rocks and tree roots. Perhaps such places are also mating sites; a number of other species of crabs are known to be able to mate only when the female has just molted.

C. hirtipes eat fresh and rotting leaves and fruit from a wide variety of Palauan plants. They feed both day and night, generally staying near their burrows during the day and retreating into them when disturbed. Around the full moon during the months in which they reproduce, they contain comparatively little and comparatively lean meat. Apparently much of the body fat is mobilized to make eggs. Palauans nonetheless catch these crabs on their seaward reproductive migrations because they like to eat the eggs. When returning from the beach the crabs are generally ignored. They tend to be fattest around the new moon. Palauans state that molting, migration, and reproduction of several other crustaceans also manifest lunar periodicity. Although the accounts I obtained were fragmentary and I did not get around to investigating them, they are of sufficient interest to merit some discussion.

Fishermen from Palau, as well as Yap, Kosrae (Kusaie), and Ponape informed me that the mangrove crab, *Scylla serrata* (*chemang*), exhibit lunar rhythms. Although this species is a valuable source of food in the Indo-West Pacific, biologists know little about its behavior in the field (Hill, 1975). It is well known from laboratory studies, however, that the male mangrove crab carries the female around beneath him for several days prior to her precopulatory molt (e.g., Ong, 1966). Palauan crab fishermen say that in the field the male finds a female on the reef, picks her up, and carries her beneath him for several days as he moves into the mangroves.

The tracks of the male crab carrying females are different from those of unencumbered males because one of the three pair of walking legs, as well as the chelipeds, is used to hold the female. (Ong's [1966] description of "doublers" is in agreement with this point.) The tracks show in mud bottoms but not in sand. They are obliterated rapidly by tidal flushing, hence a crab fisherman who sees tracks generally follows them because he knows they are probably fresh.

In the mangroves, near the inner limit of spring tidal influence, the male excavates a gently sloping hole in the mud as much as ten feet long, which he then occupies with the female for several days. Sometimes pairs of crabs will take up residence in natural crevices among mangrove roots or in holes in logs instead of digging a burrow. During this time the male "watches" as the female molts. Copulation then apparently takes place, judging by laboratory studies, but fishermen did not mention this. Solitary males also use these holes for shelter when they molt, as well as at other times. (Females will eat soft, molted males, according to fishermen, but males will not eat molted females.) Burrows may last for years during which they may be used by many different crabs.

Only after the shell has hardened, several days after molting, will the crabs leave the shelter of their holes and move back into deeper water beyond the mangroves. Here the females are often found partly buried in small sandy excavations on the fringing reef, typically in channels, or *debochel*, created geologically by stream outflow but now generally bathed in waters of high salinity. Because females moving into the mangroves do not possess eggs, it is here on the reef, Palauans surmise, that the females must "lay their eggs." (Actually the eggs hatch and the larvae are released while the eggs are still attached to the female. Prior to hatching the eggs are transferred from within the female to its swimmerets. Palauans often find the eggs inside the female, but very rarely encounter females in berry.) The apparent association of burying with ovulation fits well with the statement of Norse and Fox-Norse (1977) that "portunid crabs thus far observed cannot attach eggs unless the females bury partly in soft sediment."

The reported movement toward the sea to complete the reproductive cycle corresponds with observations made elsewhere (Ariola, 1940; Brick, 1974; Hill, 1975). After a pelagic larval period (with which Palauans are understandably unfamiliar) young crabs migrate into the mangroves. Although some breeding occurs throughout the year according to fishermen, peak breeding occurs during the spring judging by the fact that some mangrove areas swarm with juveniles from spring through early summer.

Fishermen say that the migrations of *chemang* in and out of the mangroves exhibit a lunar rhythm that seems to differ in different parts of Palau. I was unable to obtain agreement on the details, but observations in other areas support the idea that some kind of lunar periodicity is operating. Mangrove crabs in Kosrae (Kusaie) and Ponape, Micronesia, are most abundant on the reef around full moon according to fishermen from those islands (see also Bascom, 1946). Yapese fishermen similarly state that mangrove crabs "lay their eggs" on the reef around full moon. According to Vaea and Straatmans (1954), mangrove crabs in Tonga leave their burrows in the back of the mangrove at full moon. Thomas (1972) states that in Pulicat Lake, a brackish lagoon near Madras, India, mangrove crabs go *to* their burrows at the water's edge around the full moon. Heath (1971) stated that in East Africa "it is popularly supposed that more *Scylla* may be caught (in the mangroves) at the full moon," and he provided limited data that tended to support this contention. Burdon (1959) states that more mangrove crabs are caught during spring tides (i.e., around new *and* full moons) around Singapore.

What are we to make of these widespread, intriguing, but inconsistent reports? Perhaps breeding migrations into the mangroves and migrations seaward to release larvae both occur primarily during spring tide periods. In mangrove areas where there are emergent mudflats, the final portion of the inward migration and the initial portion of the seaward migration would probably *have* to occur during spring tide periods. During neap tides the crabs would have to travel some distance out of the water in order to get to and from their burrows. It is unlikely that they could do so; when they are stranded out of the water their gill leaves apparently collapse, greatly reducing the area available for oxygen exchange. Respiration drops by more than 90 percent (Veerannan, 1974), heart-rate drops by about 50 percent (Hill and Koopwitz, 1974), and they cease movement (Dickinson, 1977). The apparent conflicts in the various accounts of lunar migratory rhythms noted above could be reconciled if migration were primarily seaward around full moon and into the mangroves around new moon in some areas, whereas the sequence were reversed—that is, seaward migration around new moon, etc.—in other areas, occurring during spring tides in either case. One Palauan fisherman stated that migrations in both directions were common on the spring tides on both new and full moons.

Four different Palauan fishermen independently told me that both mangrove crabs and reef crayfish, *Panulirus* sp. (*cheraprukl*) molt predominantly on neap tides. Tahitian fishermen state that crayfish and crabs molt around the twenty-fourth day of the lunar month—that is, during one of the two neap tide periods of the month (Oliver, 1974). Burdon (1959) states that the mangrove crabs are watery-fleshed during neap tides. This likewise suggests that they molt on neap tides as the water content of crab flesh increased markedly at molting (Knowles and Carlisle, 1956). There is a conceivable advantage to timing molts to coincide with neap tides. Just-molted crustaceans are exceedingly weak and able to move only with difficulty until the new exoskeleton has hardened. It would be an advantage therefore to molt at times when tidal currents were weakest, that is, during neap tidal periods.

Palauans commonly mentioned a mangrove-dwelling land crab, *rekung el daob*, which migrated to the water's edge on sandy beaches to release its larvae several days after the new and/or full moon. There was a disagreement as to timing among my informants and I did not investigate nor identify the species.

A number of other crustaceans are known to have lunar reproductive rhythms (e.g., Reaka, 1976; Saigusa and Hidaka, 1978; Zucker, 1978).

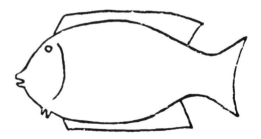

APPENDIX C

TOBIAN FISHHOOKS

English hook terminology is explained on page 113.

JABBING HOOKS

Suheriong
- Point facing upward and outward. Shank straight or almost straight.
- Used on trolling lures for tuna and for shallow dropline fishing.
- In dropline fishing with this hook the line must be jerked before the fish turns with the bait in its mouth.
- Hooks in the top of the jaw.

Yawariyet
- As with the *suheriong*, the point faces upward and outward and the shank is almost straight. But the width of the bend is less than that of the *suheriong*. This results in it being a little harder to hook a fish successfully than with a *suheriong*, but it is easier to hold the fish once hooked because of the longer point and the decreased distance between shank and point. This distance is occasionally reduced further by bending the shank toward the point.
- Designed for fish which bite aggressively, such as tuna, rainbow runners, and jacks.
- Hooks in the top of the jaw.

Hapi Sereh

- Short straight point facing upward or slightly outward.
- A versatile shallow water hook, used on trolling lures for tuna, for pole fishing on the reef, and, with a weight, for shallow dropline fishing.
- Good for large-mouthed fish.
- The fish should be allowed to turn before striking when dropline fishing.
- Hooks up badly on the bottom.

Sangi

- Named after a village in Halmahera, Moluccas, from where it was reportedly introduced centuries ago by a fisherman who was blown off course and drifted to Tobi.
- Narrow, U-shaped bend, long, straight point directed very slightly outward from shank. Shank straight or gently curved.
- Used with shell lures for tuna, and in smaller sizes with feathers for small-mouthed squirrelfish (e.g., *Holocentrus* sp.) and other small-mouthed fishes. Trolled slowly.

Fotomahech

- Similar in shape to *sangi*, but with wider bend, and point directed slightly inward. Shank occasionally bent inward when used in trolling with the lure some distance from the canoe.
- Used with feather lure in slow trolling for tuna, and in smaller sizes for wide-mouthed fish, such as most squirrelfish. Also used in shallow dropline fishing.
- The fish should be allowed to turn with the hook in its mouth before striking.
- Large fish tended to bend the point of the turtle shell version outward, allowing them to escape.
- The fish must be allowed to turn away with the hook in its mouth before striking.
- Hooks in the side of the mouth.

Man Tanante

- Shank and point are one continuous curve. Shank often very short. Point directed toward top of shank.
- Most versatile shallow water dropline and reef pole fishing hook, used when "you don't know what you are going to catch."
- The fish should be allowed to turn before striking.
- Hooks in the side of the mouth.

Fichifichino'o

- The name means "bent like a palm leaflet midrib." When the latter is bent the stiff core breaks, creating a sharp angle rather than a smooth bend. The hook has three (or occasionally four) such angular bends in it. The point faces the top of the shank.
- Used for trolling with a lure, and with a weight for shallow dropline fishing.
- Less liable to break under the strain of a fish than other turtle shell hooks. Less liable to bend than other metal hooks. Small hooks of this design can effectively hook large fish.
- The fish should be allowed to turn with the bait in its mouth before striking.
- Hooks in the side of the mouth.

ROTATING HOOKS

Metch

- The name is derived from the Tobian name for cone shells, and refers to the spiral groove on the crown of the shell; the lower shank, bend, and point form a spiral.
- Strongly curved point facing inwards (shallower water version) or downwards (deeper water version).
- Used with a weight in dropline fishing for large-mouthed fish such as grouper. This is the hook of choice when fishing very deep.
- The fish must be allowed to run with the hook and then retrieved while keeping steady moderate tension on the line (see page 114).

Fahum

- Bend almost circular.
- A deep water dropline hook used for fish with smaller mouths (e.g., lethrinids) than those sought with a *metch*.
- As many as five may be baited and attached to a single line at intervals.
- Used without a weight (a rock is tied loosely to the line to take the hook down, then shaken loose with a jerk).
- The hooking technique is intermediate between that used for a conventional rotating hook and a jabbing hook (see chap. 9). The fish is played gently for a while until the hook sets itself lightly. Then the line is jerked hard to set the hook more firmly.

Turtle shell
version

Ramatiho
- Similar in shape to *fahum*, but point does not approach the shank so closely and the shank has more inward curvature.
- Used for dropline fishing in fairly deep water.
- The bait is placed on the hook loosely to allow maximum penetration; when a fish bites the bait usually slides up the shank.
- Used for groupers and also for fish that suck the bait in and out of their mouths cautiously; the wide bend makes the hook hard to spit out.
- Always hooks in the jaws, not in the lips; good for fish with easily torn lips.
- A strong fish is liable to bend this hook and escape.

FONG HOOKS

"Old style"

"New style"

Haufong
- Straight point. Shank straight or angled inward for use in deeper water. Point, except for *fong*, leaning outward from shank. Made only of metal.
- Used for dropline fishing.
- Good for most small-mouthed carnivorous fish, but specifically designed for catching triggerfish (see p. 117, 118).
- A small piece of bait, which just covers the *fong*, is used.

Ramaributs
- Shank straight, or angled inward for use in deeper water. Point, limb except for *fong*, leaning outward from shank. Turtle shell version has both barb and *fong*.
- Used in dropline fishing in shallow to moderately deep (eighty fathoms) water for tuna and "slow-biting" fish. Used with a weight. Live or dead goatfish often used for bait.
- The *fong*, in addition to helping keep the bait (and the fish) on the line, causes the hook to perform like a rotating hook (see p. 113). The fish is allowed to swim with the bait, then retrieved with a steady pull.
- A strong fish is liable to snap off the *fong*.

Ramasu
- Similar in shape to *haufong*, but the fong is longer and the bait is placed below rather than around it.
- In the past, before the invention of the *haufong*, this hook was used for triggerfish. It is seldom used today.

GLOSSARY
OF PALAUAN WORDS

Bad El Wel "Sleeps like a turtle"
Bebael Rabbitfish, *Siganus punctatus*
Bedaoch Black noddy, *Anous tenuirostris*
Beduut Rabbitfish, *Siganus argenteus*
Belai Blue-lined surgeonfish, *Acanthurus lineatus*
Belau Palau
Besechaml Large-eyed porgy, *Monotaxis grandoculis*
Bidekill Cast net
Bikl Sweetlips, *Plectorhynchus obscurus*
Biturchturch Double-line or shark-mackerel, *Grammatorcynus bicarinatus*
Brer Bamboo raft
Bsukl Squirrelfish of the genus *Myripristis*
Budech Yellow-cheeked wrasse, *Choerodon anchorago*
Bul Restriction, conservation regulation
Butiliang Generic name for green parrotfish
Chai Great barracuda, *Sphyraena barracuda*
Chaol (or Chaol Diong) The young of the milkfish, *Chanos chanos*
Chederobk (also Cherobk) Big-head jack, *Caranx ignobilis*
Chedochd Moharra, *Gerres abbreviatus*
Chedui Blue-lined sea bream, *Symphorus spilurus*

Chei The waters between the shore and the outer reef slope
Chelas Convict surgeonfish, *Acanthurus triostegus*
Chemang Mangrove crab, *Scylla serrata*
Chemisech Samoan goatfish, *Mulloidichthys flavolineatus*
Cheraprukl Spiny lobster, *Panulirus* sp
Chersuuch Dolphinfish, *Coryphaena hippurus*
Chesau Mullet, *Chelon vaigiensis*
Chesengel Surgeonfish, *Acanthurus mata*
Choas Millipede, *Polyconoceras callosus*; Sea cucumber, *Holothuria atra*
Chudch A snapper, *Lethrinus ramak*
Chudel Green jobfish, *Aprion virescens*
Chum Unicornfish, *Naso unicornis*
Dech Samoan goatfish, *Mulloidichthys flavolineatus*
Desachel Squirrelfish, *Adioryx spinifer*
Desui Rainbow runner, *Elagatis bipinnulatus*
Dub Fish poison made from *Derris* root
Dukl Triggerfish, *Pseudobalistes flavimarginatus*
Eluikl Emperor, *Lethrinus* sp
Iab Gold-spotted jack, *Carangoides fulvoguttatus*
Iengel Line on fish stringer
Itotch Thumb-print emperor, *Lethrinus harak*
Ius Crocodile, *Crocodylus porosus*
Katsuo Skipjack tuna, *Katsuwonus pelamis* (from Japanese)
Kedesau Red snapper, *Lutjanus bohar*
Kedesauliengel Mangrove snapper, *Lutjanus argentimaculatus*
Kelat Mullet, *Crenimugil crenilabus*
Kemedukl Bumphead parrotfish, *Bolbometopon muricatus*
Keremlal Paddle-tail snapper, *Lutjanus gibbus*
Kesebekuu Generic name for moray eels
Keskas Wahoo, *Acanthocybium solandri*
Kesokes Stationary net used to trip fish on a falling tide
Kitlel A freshwater eel, *Anguilla marmorata*
Klibiskang Single pronged throwing spear
Klsbuul Rabbitfish, *Siganus lineatus*
Klspeached Yellowtail scad, *Caranx mate*
Komud Rudderfish, *Kyphosus cinerascens* and *K. vaigiensis*
Kotikw Moharra, *Gerres oblongus*
Ksous Stingray, *Dasyatus sephen*
Kutalchelbeab Unidentified ray
Lmall The period when a number of species of parrotfish are said to spawn;
that is, the sixth to the tenth day of the lunar month
Mamel Bumphead wrasse, *Cheilinus undulatus*
Manguro Yellowfin tuna, *Thunnus albacares* (from Japanese)
Matukeyoll Blacktip shark, *Carcharhinus melanopterus*
Meai Barracuda, *Sphyraena forsteri* and *S. genie*
Meas Rabbitfish, *Siganus canaliculatus*
Mechur Small-toothed emperor, *Lethrinus microdon*
Mederart Grey reef shark, *Carcharhinus amblyrhynchos*

Mekebud   Herring, *Herklotsichthys punctatus*
Mengai   Ring-tailed unicornfish, *Naso annulatus*
Mengeai   Neap tide period, the days around the half moon.
Mengerenger   Banded sea snake, *Laticauda colubrina*
Mengeslbad   A snapper, *Lutjanus semicinctus*
Mersad   Wahoo, *Acanthocybium solandri*
Mersaod   Great barracuda, *Sphyraena barracuda*
Mesekelat   *Chanos chanos*, milkfish
Mesekiu   Dugong, *Dugong dugon*
Mesekuuk   Ring-tailed surgeonfish, *Acanthurus xanthopterus*
Mesikt   The Pleiades
Metal   Lemon shark, *Negaprion acutidens*
Metngui   Small-toothed jobfish, *Aphareus rutilans*
Miich   Tree, *Terminalia catappa*
Mirorch   Grouper, *Ephinephelus merra*
Mlangmud   Long-nosed emperor, *Lethrinus miniatus*
Mokas   Grouper, *Plectropomus leopardus*
Mokorkor   Double line or shark mackerel, *Grammatorcynus bicarinatus;* also called *biturchturch*
Mordubch   Greater barracuda, *Sphyraena barracuda*
Ngebard   West; the season of westerly winds (roughly from May through October)
Ngelngal   Spanish mackerel, *Scomberomorus commersoni*
Ngemngumk   Angelfish
Ochaieu   Audubon's shearwater, *Puffinus lherminieri;* eagle ray, *Aetobatus narinari*
Odoim   Animal food
Ongos   East; the season of easterly winds (roughly from November through April)
Oruidl   Jack, *Caranx melampygus*
Osel   Net used in conjunction with a *ruul*
Oungeuaol   Offshore shark fishing
Plutek   Aggressive, rapidly moving school of sharks
Rams   Sand channel leading through a reef flat to the reef edge
Rekung El Beab   Land crab, *Cardisoma hirtipes*
Rekung el Daob   Unidentified land crab
Remochel   Grouper, *Epinephelus fuscoguttatus*
Ruul   Leaf sweep
Sebus L'Toach   An estuarine cardinalfish
Sekos   Generic name for needlefish
Semael   Wooden fish trap used in conjunction with a *ruul*
Seosech   White tern, *Gygis alba*
Soda (or Soda Katsuo)   Mackerel tuna, *Euthynnus affinis* (from Japanese)
Suld   Bonefish, *Albula vulpes*
Taod   Multiple-pronged throwing spear
Tebukbuk   Unidentified ray
Tekoi L'Chei   "Words of the lagoon"
Terekrik   Ox-eye scad, *Trachurus boops*

Tkuu   Yellowfin tuna, *Thunnus albacares*

Tmur Ra Ongos   The first lunar month of the season of ongos

Tukidolch   "Tide gauge" eel

Udech   Emperor, *Lethrinus ramak*

Ulach   Cornetfish, *Fistularia petimba;* also generic name for hammerhead shark

Ulekiued   Fusiliers, *Caesio* sp. and *Pterocaesio* sp.

Uloi   Archerfish, *Toxotes jaculatrix*

Ulupsuchl   Reef white tip shark, *Triaenodon obesus*

Uluu   Mullet, *Chelon vaigiensis*

Urur   A mangrove tree, *Sonneratia alba*

Wii   Golden jack, *Gnathanodon speciosus*

Yaos   Diagonal-banded sweetlips, *Plectorhynchus goldmanni*

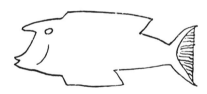

GLOSSARY
OF TOBIAN WORDS

Arm The narrow, rough stream of water sometimes extending downstream from either side of an island
Bub Southern Cross; also generic name for triggerfish
Bucho Triggerfish, *Balistapus undulatus*
Bwerre Certain species of groupers (serranids); also refers to a multihook dropline fishing technique
Chariferigut Cleaner wrasse, *Labroides dimidiatus*
Chera Generic name for remoras
Cherouchouko Certain species of jacks (carangids)
Doriyout Sonsorolese name for *suriyout*
Echarivus Lemon shark, *Negaprion acutidens*
Fahum A type of Tobian fishhook
Farupon Pisahe He Pulo Blue-banded angelfish, *Pygoplites diacanthus*
Faumer Tripletail, *Lobotes surinamensis*
Fen A baited hoop net
Fichifichino'o A type of Tobian fishhook
Fong The inturned tip on certain Tobian fishhooks
Fotomahech A type of Tobian fishhook
Fotorimar Sea perch, *Anthias huchti*
Hafira A species of snapper
Hao Parrotfish, *Cetoscarus pulchellus* and *Chlorurus stronglycephalus*

Hapi Sereh   A type of Tobian fishhook

Hapitsetse   The area of very rough water where two currents converge downstream of an island

Hari   Generic term for bright-colored groupers

Hari-Merong   Blue-spotted grouper, *Cephalopholis cyanostigma*

Hasetiho   The near-island current patterns associated with a stable eddy pair

Hatih   Generic name for dull-colored snappers

Haufong   A type of Tobian fishhook

Haugus   A species of grouper

Hob   Angelfish, *Centropyge bicolor*

Horach   Convict surgeonfish, *Acanthurus triostegus*; also the name of a tree

Machuyar   A species of parrotfish (scarid)

Mahi   Generic name for needlefish

Mam   Bumphead wrasse, *Cheilinus undulatus*

Man Tanante   A type of Tobian fishhook

Martacham   A species of squirrelfish

Marusetih   Master fisherman

Masuh   A species of parrotfish (scarid)

Matirai   Bumphead parrotfish, *Bolbometopon muricatus*

Mecheri   A species of wrasse

Merangu   A species of unicornfish, *Naso* sp.

Merifuts   A species of barracuda, *Sphyraena* sp.

Metch   A type of Tobian fishhook

Moghu   Surgeonfish, *Acanthurus glaucopareius*

Moharechoh   Sailfish, *Istiophorus platypterus*

Mor   Generic name for squirrelfish

Neihahi   The "peaceful" season of gentle winds and calm water, from February through September

Ngungpaha   Imperial angelfish, *Pomacanthus imperator*

Ngusngus   A species of squirrelfish

Niyafang   The season of rough water, from October through January

Paho   Generic name for sharks

Paip   Generic name for butterflyfish (chaetodontids)

Pingao   Sailfin tang, *Zebrasoma veliferum*

Ramaributs   A type of Tobian fishhook

Ramasu   A type of Tobian fishhook

Ramatiho   A type of Tobian fishhook

Rau   Harlequin sweetlips, *Plectorhynchus chaetodontoides*

Reebai   Either of two species of small-mouthed wrasse

Reyu   Generic name for stonefish and other scorpaenids

Richoh   Generic name for damselfish (pomacentrids) and small snappers (lutjanids)

Richopah   A species of damselfish (pomacentrid)

Riper Yecho   Coconut meat; also those species of damselfish (pomacentrids) that are caught using coconut meat for bait

Riwesiri Yapetase   Small dolphinfish, *Coryphaena hippurus* (Sonsorolese)
Ruhuruho   Old generic name for triggerfish
Sabacho   Nurse shark, *Ginglymostoma ferrugineum*
Sangi   A type of Tobian fishhook
So'owo   Goatfish
Suchowa   A barracuda, *Sphyraena* sp.
Suheriong   A type of Tobian fishhook
Suriyout   Area of calm water upstream of an island
Tanganangan   A small species of wrasse
Tauchacha   Marlin
Tavitef   Generic name for boxfish
Temaubour   Probably a spiny-cheeked perch (grammistid)
Teribour   A sweetlips (pomadaysiid)
Teter   A jack (carangid)
Urech   False moorish idol, *Heniochus acuminatus*
Uremar   Certain sweetlips (pomadasyids) with black markings
Usurifarema   A type of frogfish
Wasori   Moorish idol, *Zanclus cornutus*
Watur   A deep-dwelling emperor
Yar   Wahoo, *Acanthocybium solandri*
Yassur   Medium-sized dogtooth tuna, *Gymnosarda unicolor*
Yawariyet   Type of Tobian fishhook
Yetam   Pilot fish, *Naucrates ductor*

REFERENCES CITED

Abe, M. 1938. Palau plants used for fishing. Zool. Mag. (Tokyo) 50:44. (In Japanese.)

Acheson, J. A. 1972. Territories of the lobstermen. Natural History 81:60–69.

Akimichi, T. 1978. The ecological aspect of Lau (Solomon Islands) ethnoichthyology. J. Polyn. Soc. 87:301–326.

Akoev, G. N., O. B. Ilynsky, and P. M. Zadan. 1976. Responses of electroreceptors (Ampullae of Lorenzini) of skates to electric and magnetic fields. J. Comp. Physiol. 106:127–136.

Aleev, Y. G. 1969. Function and gross morphology in fish. (Translated from Russian.) Keter Press, Jerusalem.

Alexander, P. 1975. Do fisheries experts aid fisheries development? Marit. Stud. Mgmt. 3:5–11.

Allen, G. R., D. F. Hoese, J. R. Paxton, J. E. Randall, W. A. Starck, F. H. Talbot, & G. P. Whitley. 1976. Annotated checklist of the fishes of Lord Howe Island. Rec. Australian Mus. 30:365–454.

Allen, J. A. 1966. The rhythms and population dynamics of decapod crustaceans. Oceanogr. Mar. Biol. Ann. Rev. 4:247–265.

Anderson, E. 1963. Tahitian bonito fishing. Kroeber Anthrop. Soc. Pap. 28:87–120.

Anell, B. 1955. Contribution to the history of fishing in the southern seas. Studia Ethnographica Upsaliensia 9.

Ariola, F. J. 1940. A preliminary study of the life history of Scylla serrata (Forskål). Philip. J. Sci. 73:437–456.

Ashmole, N. P., and M. J. Ashmole. 1967. Comparative feeding ecology of sea birds of a tropical ocean island. Peabody Mus. Nat. Hist., Yale Univ. Bull. 24.

Atema, J. 1980. Chemical senses, chemical signals, and feeding behavior in fishes, *in* J. Bardach, J. J. Magnuson, R. May, and J. Reinhart (eds.) Fish behavior and its use in the capture and culture of fishes. I.C.L.A.R.M., Manila.

Bagnis, R., P. Mazellier, J. Bennett, and E. Christian. 1972. Fishes of Polynesia. Les Editions du Pacifique Papeete, Tahiti.

Baldwin, W. J. 1977. A review of the use of live bait fishes to capture skipjack tuna, *Katsuwonus pelamis*, in the tropical Pacific Ocean with emphasis on their behavior, survival and availability. Nat. Ocean. Atmosph. Admin. Report. Nat. Marine Fish. Serv. Circ. 408:8−35.

Balfour, H. 1913. Kite fishing. Pp. 583−607, *in* Essays and Studies Presented to Wm. Ridgeway. Cambridge Univ. Press.

Banner, A. 1972. Use of sound in predation by young lemon sharks, *Negaprion brevirostrus* (Poey). Bull Mar. Sci. 22:251−283.

Banner, A. H. 1977. Hazardous marine animals, Pp. 1378−1436 *in* C. G. Tedeschi, W. G. Eckert, and L. G. Tedeschi (eds.) Forensic medicine: a study in trauma and environmental hazards. Vol. III. Environmental Hazards. W. B. Saunders Co.

Baranov, F. I. 1976. Selected works on fishing gear. Vol. I. Commercial fishing techniques. (Translated from Russian by Israel Translation Service.) Keter Publishing House, Ltd. Jerusalem.

Barry, M. 1978. Behavioral response of yellowfin tuna, *Thunnus albacares*, and kava kava, *Euthynnus affinis* to turbidity. Dissertation, Univ. of Hawaii.

Bascom, W. R. 1946. Ponape: A Pacific economy in transition. U.S. Commercial Company. Economic Survey of Micronesia. Vol. 8.

Batchelor, G. K. 1967. An Introduction to Fluid Dynamics. Cambridge Univ. Press.

Bates, M., and D. Abbott. 1958. Coral Island: Portrait of an Atoll. Scribner's.

Beaglehole, E., and P. Beaglehole, 1938. Ethnology of Pukapuka. Bishop Mus. Bull. 150.

Beasley, H. G. 1928. Pacific island records, fish hooks. Seeley Service, London.

Beck. H. 1973. Folk-lore and the sea. Wesleyan Univ. Press.

Becke, L. 1905. Notes from my South Seas Log. London.

Bergman, R. 1970. *In* E. C. Janes (ed.) Fishing with Ray Bergman. Knopf.

Berlin, B., D. E. Breedlove, and P. H. Raven. 1974. Principles of Tzeltal plant classification. Academic Press.

Best, E. 1929. Fishing methods and devices of the Maori. Dominion Mus. (N.Z.) Bull. 12.

Billings, V. and J. Munro. 1974. The biology ecology and bionomics of Caribbean reef fishes. Part Ve. Pomadaysyidae (grunts). Res. Rept. Zool. Dept., Univ. West Indies. No. 3.

Black, P. W. 1968. The fishing lore of Tobi. Unpublished report. (A copy is located at the Palau Museum.)

————. 1977. Neo-Tobian culture: Modern life on a Micronesian atoll. Ph.D dissertation, Univ. of California, San Diego.

Blurton Jones, N., and M. J. Konner, 1976. !Kung knowledge of animal behavior (or: The proper study of mankind is animals). Pp. 325–348 *in* R. B. Lee and I. DeVore (eds.) Kalahari Hunter-Gatherers. Harvard Univ. Press.

Boyd, F. D., J. C. Dunlop, A. L. Gillespie, G. Gulland, E.D.W. Grieg, S. C. Mahalanobis, M. I. Newbigin, and D. N. Paton. 1898. The physiology of salmon in fresh water. J. Physiol. 22:333–356.

Breder, C. M. Jr. 1959. Observations on the spawning behavior and egg development of *Strongylura notata* (Poey). Zoologica 44, Part 4, 141–148.

Breder, C. M. Jr. and D. E. Rosen. 1966. Modes of reproduction in fishes. Natural History Press.

Brick, R. W. 1974. Effects of water quality, antibiotics, phytoplankton and food on survival and development of larvae of *Scylla serrata*. Aquaculture 3:231–244.

Brigham, W.T.A. 1903. A handbook for visitors to the museum. B.P. Bishop Mus. Spec. Publ. 3.

Bromhall, J. D. 1954. Notes on the reproduction of the grey mullet, *Mugil cephalus* Linnaeus. Hong Kong Univ. Fish. J. 1:19–34.

Brower, K. 1974. With their islands around them. Holt, Rinehart & Winston.

Brown, G. 1910. Melanesians and Polynesians. MacMillan & Co., London.

Bryan, P. G., and D. B. McConnell, 1976. Status of giant clam stocks (Tridacnidae) and Helen Reef, Palau, Western Caroline Islands, April 1975. Mar. Fish. Rev. 38:15–18.

Buck, P. H. 1930. Samoan material culture. B.P. Bishop Mus. Bull. 75.

————. 1932. Ethnology of Manihiki and Rakahanga. B.P. Bishop Mus Bull. 99.

————. 1949. The coming of the Maori. Whitcombe and Tombs, Christchurch, N.Z.

Bulmer, R.H.H., J. I. Menzies, and R. Parker. 1975. Karam classification of reptiles and fishes. J. Polyn. Soc. 84:267–308.

Burdon, T. W. 1959. The fishing gear of Singapore. Fish. Survey Rept. #2. Fish. Bull. #3. Singapore Govt. Printing Office.

Burnette-Herkes, J. N. 1975. Contribution to the biology of the red hind, *Epinephelus guttatus*, a commercially important serranid fish from the tropical western Atlantic. Ph.D. dissertation, Univ. of Miami.

Callaghan, P. 1976. Employment and factor productivity in the Palau skipjack fishery: a production function analysis. Ph.D. dissertation, Univ. of Hawaii.

Carr, A. 1972., Great reptiles, great enigmas. Audubon 1972, March:24–34.

Caspers. H. 1961. Beobachtungen über Lebensraum und Schwärm periodizität des Palolo wurmes, *Eunice viridis* (Polychaeta, Eunicidae). Int. Revue Ges. Hydrobiol. 46:175–183.

Catala, R. 1957. Report on the Gilbert Islands: some aspects of human ecology. Atoll Res. Bull. 59:1–187.

Chacko, P. I. 1950. Marine plankton from waters around the Krusadai Island. Proc. Indian Acad. Sci. 31:162–173.

Christy, F. T., Jr. 1969. Fisheries goals and the rights of property. Trans. Amer. Fish. Soc. 98:369–378.

Clark, E. 1953. Lady with a spear. Harper and Bros.

Clarke, T. A. 1971. The ecology of the scalloped hammerhead shark, *Sphyrna lewini*, in Hawaii. Pac. Sci. 25:133–144.

Codrington, R. H. 1891. The Melanesians: studies in their anthropology and folk-lore. Clarendon Press.

Cohen, D. 1970. How many recent fishes are there? Proc. Calif. Acad. Sci. 38:341–346.

Coles, R. J. 1910. Observations on the habits and distribution of certain fishes taken on the coast of North Carolina. Bull. Amer. Mus. Nat. Hist. 28:337–348.

Colin, P. L., and I. E. Clavijo. 1978. Mass spawning of the spotted goatfish, *Pseudopeneus maculatus* (Bloch) (Pisces : Mullidae). Bull, Mar. Sci. 28:780–782.

Cook, J. 1785. A voyage to the Pacific Ocean undertaken by command of His Majesty, for making discoveries in the Northern Hemisphere. Vol. 3.

Copley, H., and T. E. Allfree. 1950. The sea fisheries of Kenya. East African Agric. J. 15:180–186.

Cordell, J. C. 1973. Modernization and marginality. Oceanus 17:28–33.

———. 1974. The lunar-tide fishing cycle in northeastern Brazil. Ethnology 13:379–392.

Corney, B. G. 1922. Abstract of a paper on the periodicity of the swarming of Palolo (*Eunice viridis* Gr.). J. Torquay Nat. Hist. Soc. 3:126–130.

Craig, A. K. 1966. Geography of fishing in British Honduras and adjacent coastal waters. Louisiana State Univ. Stud. Coastal Stud. Ser. 14:1–143.

Cresswell, G. 1979. Appearance of tropical fish off N.S.W. coast explained. Australian Fish. 38:42–43.

Creutzberg, F. 1975. Orientation in space: invertebrates. Pp. 555–656 *in* O. Kinne (ed.). Marine Ecology Vol. 2 Pt. 2. Wiley & Sons.

Cross, R. 1978. Fisheries Research Notes (Sept. 1976–October 1977). Fisheries Division, Ministry of Commerce and Industry. Tarawa. (mimeo).

Crutchfield, J. A. 1973. Resources from the sea. Pp. 105–133 *in* T. S. English (ed.) Ocean Resources in Public Policy. Univ. Washington Press.

Dan, K., and H. Kubota. 1960. Data on the spawning of *Comanthus japonica* between 1937 and 1955. Embryologia 5:21–37.

Danielsson, B. 1956. Work and life on Raroia. Allen & Unwin.

Defant, A. 1961. Physical oceanography. Vol. 1. Pergamon Press.

Delsman, H. C. 1929. Fish eggs and larvae from the Java Sea. Treubia 11:275–286.

Diaz, D.J.V. 1884. Corrida y Arribazón de Algunos Peces Cubanos. Manuel Gomez de la Maza, Havana. 15p.

Dickie, L. M. 1962. Effects of fishery regulations on the catch of fish. Pp. 100–13 *in* R. Hamlisch (ed.) Economic Effects of Fishery Regulation. FAO, Rome.

Dickinson, R. E. 1977. The occurrence and natural habitat of the mangrove crab, *Scylla serrata* (Forskål), on Ponape and Guam. M. Sc. thesis, Univ. of Guam.

Dooley, J. K. 1972. Fishes associated with pelagic *Sargassum* complex, with a discussion of the *Sargassum* community. Contr. Marine Sci. 16:1–32.

Durkheim, E., and M. Mauss. 1903. De quelques formes primitives de classification: contribution à l'étude des representations collectives. Année Socioloqique 6:1–72.

Ehrlich, P. R. 1975. The population biology of coral reef fish. Ann. Rev. Ecol. Syst. 6:211–248.

Eibl-Eibesfeldt, I. 1964. Land of a thousand atolls. World Publ. Co.

Eilers, A. 1936. Westkarolinen: Tobi und Ngulu. Ergebnisse der Sudsee—expedition 1908–1910 (IIB9, Part I). G. Thilenius. ed. Hamburg. Friedrichsen, den Gruyter & Co.

Ellis, W. 1859. Polynesian Research. 2d ed. Vol. I H. G. Bohn, London.

Emery, A. R. 1972. Eddy formation from an oceanic island: ecological effects. Carib. J. Sci. 12:121–128.

Emory, K. P. 1975. Material culture of the Tuamotu Archipelago. Pac. Anthrop. Rec. 22:1–253.

Erdman, D. S. 1976. Spawning patterns of fishes from the northeast Caribbean. Agric. Fish. Contrib. (Dept. Agric. Puerto Rico). 8(2):1–36.

Faulkner, D. 1974. This living reef. Quadrangle/The New York Times Book Co.

Feddern, H. A. 1965. The spawning, growth, and general behavior of the bluehead wrasse, *Thalassoma bifasciatum* (Pisces: Labridae) Bull. Mar. Sci. 15:896–941.

Force, M. T. 1976. The persistence of precolonial exchange patterns in Palau: a study of cultural continuities. Ph.D. dissertation, Walden University.

Force, R. W., and M. Force. 1972. Just one house: a description and analysis of kinship in the Palau Islands. B. P. Bishop Museum Bull. 235.

Forman, S. 1967. Cognition and catch: the location of fishing spots in a Brazilian coastal village. Ethnology 6:416–426.

———. 1970. The raft fishermen: tradition and change in the Brazilian peasant economy. Indiana Univ. Press.

Forster G. R. 1973. Line fishing on the continental slope. The selective effect of different hook patterns. J. Mar. Biol. Ass. U.K. 53:749–751.

Fosberg, F. R. 1972. Man's effects on island ecosystems. Pp. 869–880 *in* M. T. Farvar and J. P. Milton (eds.) The Careless Technology. Natural History Press.

Fourmanoir, P., and P. Labout. 1976. Poissons de Nouvelle Calédonie et des Nouvelles Hébrides. Les Editions du Pacifique, Papeete, Tahiti.

Frey, D. 1951. The use of sea-cucumbers in poisoning fishes. Copeia 1951:175–176.

Fricke, H. 1971. Fische als Feinde tropischer Seeigel. Mar. Biol. 9:328–38.

Gaigo, B. 1977. Present day fishing practices in Tatana Village. Pp. 176–181 *in* J. H. Winslow (ed.) The Melanesian Environment. Australian Nat. Univ. Press, Canberra.

Gatty, H. 1953. The use of fish poison plants in the Pacific. Trans. Proc. Fiji Soc. Sci. Industry 3:152–159.

Gifford, C. A. 1962. Some observations on the general biology of the land crab *Cardisoma quanhumi* (Latreille), in South Florida. Biol. Bull. 123:207–233.

Gill, T. N. 1909. The archer-fish and its feats. Smithsonian Misc. Coll. 52:277–286.

Ginetz, R. M., and P. A. Larkin. 1973. Choice of colors of food items by rainbow trout *(Salmo gairdneri)* J. Fish. Res. Bd. Canada 30:229–234.

Girschner, M. 1912. Die Karolineninsul Namoluk und ihre Bewohner. Baessler-Archiv. 2:123–215.

Gladwin, T. 1970. East is a big bird. Harvard Univ. Press.

Glynn, P. W. 1973. Ecology of a Caribbean coral reef. The *Porites* reef flat biotope: Part II. Plankton community with evidence for depletion. Mar. Biol. 22:1–22.

Goeden, G. B. 1978. A monograph of the coral trout *Plectropomus leopardus Lacépède.* Queensland Fish. Serv. Res. Bull. No. 1:1–42.

Gooding, R. M., and J. J. Magnuson. 1967. Ecological significance of a drifting object to pelagic fishes. Pac-Sci. 21:486–497.

Gordon, H. S. 1954. The economic theory of a common property resource: the fishery. J. Political Econ. 62: 124–142.

Gosline, W. A., and V. E. Brock. 1960. Handbook of Hawaiian Fishes. Advertiser Publishing Co.

Grimble, A. 1931. Gilbertese astronomy and astronomical observations. J. Polyn. Soc. 40:197–224.

Groves, W. C. 1933–34. Fishing rights at Tabar. Oceania 4:432–457.

Günther, W. 1938. Bestehen zusammenhänge zwischen Geburtstermin, Geschlecht des Kindes und Mondstellung. Zbl. Gynäk 1938: 1196–1202.

Guppy, H. B. 1887. The Solomon Islands and their natives. London.

Hallier, J. P. 1977. Artisanal fisheries in the New Hebrides: the Lamap fishing school. South Pacific Comm. Fisheries Newsletter 14:8–11.

Halstead, B. W. 1978. Poisonous and Venomous Marine Animals of the World. Darwin Press, Inc.

Hamner, W. M., and I. Hauri. 1977. Fine-scale surface currents in the Whitsunday Islands, Queensland, Australia: effect of tide and topography. Austr. J. Mar. Freshw. Res. 28:333–359.

Handy, E. S. C. 1932. Houses, boats and fishing in the Society Islands. Bishop Mus. Bull. 90.

Handy, E. S. C., E. G. Handy, and M. K. Pukui. 1972. Native planters in old Hawaii. Bishop Mus. Bull. 233.

Hankin, J. H., and L. E. Dickinson. 1972. Urbanization, diet and potential health effects in Palau. Amer. J. Clinical Nutrit. 25:348–353.

Hanlon, J. 1979. When the scientist meets the medicine men. Nature 279:284–285.

Harden-Jones, F. R. 1968. Fish migration. Edward Arnold, London.

Hardin, G. 1968. The tragedy of the commons. Science 162:1243–1248.

Harley, T. 1970. Moon Lore. C. E. Tuttle.

Harmelin-Vivien, M. L. 1977. Ecological distribution of fishes on the outer slope of Tulear Reef (Madagascar). Pp. 290–295. Proc. 3rd Internat. Coral Reef Symp. Vol. I.

Harrington, R. W. Jr. 1947. Observations on the breeding habits of the yellow perch, *Perca flavescens* (Mitchill). Copeia 1947:199–200.

Harry, R. R. 1953. Ichthyological field data of Raroia Atoll, Tuamotu Archipelago. Atoll Res. Bull. 18:1–190.

Hasse, J. J., B. B. Madraisau, and J. P. McVey. 1977. Some aspects of the life history of *Siganus canaliculatus* (Park) (Picses: Siganidae) in Palau. Micronesica 13:297–312.

Hattori, S. 1970. Preliminary note on the structure of the Kuroshio from the biological point of view, with special reference to pelagic fish larvae. Pp. 399–446 *in* J. C. Marr (ed.) The Kuroshio: A Symposium on the Japan Current. East-West Center Press, Honolulu.

Heath, J. R. 1971. Some preliminary results of tropical fishing trials for crabs. East African Agric. & Forestry J. Oct. 1971:142–145.

Helfman, G. 1968. Disturbing the hawksbill gently. Micrones. Reporter 16(3): 10–12.

Helfman, G. S. 1978. Patterns of community structure in fishes: summary and overview. Env. Biol. Fish. 3:129–148.

Helfman, G. S., and J. E. Randall. 1973. Palauan fish names. Pac. Sci. 27:136–153.

Helfrich, P., and P. M. Allen. 1975. Observations on the spawning of mullet *Crenimugil crenilabis* (Forskål), at Enewetak, Marshall Islands. Micronesica 11:219–225.

Hidikata, H. 1942. Calendar of Palau natives. South Sea Is. 8(1):14–24 (in Japanese).

High, W. L., and I. E. Ellis. 1973. Underwater observations of fish behavior in traps. Helgol. Wiss. Meeres. 24:341–347.

Hill, B. J. 1975. Abundance, breeding and growth of the crab *Scylla serrata* in two South African estuaries. Mar. Biol. 32:119–126.

Hill, B. J., and H. Koopwitz. 1974. Heart-rate of the crab. *Scylla serrata* in air and in hypoxic conditions. Comp. Biochem. Physiol. 52A:385–387.

Hill, H. B. 1978. The use of nearshore marine life as a food resource by American Samoans. Pacific Islands Progr., Univ. of Hawaii. Misc. working pap. 1978:1, 170p.

Hind, V. T. 1969. A fisheries reconnaissance to Wallis Island. South Pacific Commission. MS.

Hisard, P., Y. Magnier, and B. Wauthy. 1969. Comparison of the hydrographic structure of equatorial waters north of New Guinea and at 170°E. J. Mar. Res. 27:191–205.

Hobson, E. S. 1963. Feeding behavior in three species of sharks. Pac. Sci. 17:171–194.

———. 1965. Diurnal–nocturnal activity of some inshore fishes in the Gulf of California. Copeia 1965:291–302.

———. 1966. Visual orientation and feeding in seals and sea lions. Nature 210(5033):326–327.

———. 1972a. Activity of Hawaiian reef fishes during the evening and morning transitions between daylight and darkness. Fish. Bull 70: 715–740.

———. 1972b. The survival of Guadalupe cardinalfish *Apogon guadalupensis* at San Clemente Island. Calif. Fish and Game 58:68–69.

———. 1973. Diel feeding migrations in tropical reef fishes. Helgol. Wiss. Meeresunters. 24:361–370.

Hocart, A. M. 1929. Lau Islands, Fiji. B. P. Bishop Mus. Bull. 62.

Hodgson, E. S. 1978. Knowledge and exploitation of the sensory biology of sharks in the southwestern Pacific. Pp. 57–67 *in* E. S. Hodgson and R. Mathewson (eds.) Sensory biology of sharks, skates and rays. Office of Naval Res., Dept. of the Navy, Arlington, Virginia.

Holden, H. 1836. A narrative of the shipwreck, captivity and suffering of Horace Holden and Benjamin Nute. Weeks, Jordan and Co., Boston.

Hoyt, E. 1956. The impact of a money economy on consumption patterns. Ann. Amer. Acad. Pol. Soc. Sci. 303:12–22.

Hsiao, S. C., and A. L. Tester. 1955. Reaction of tuna to stimuli, 1952–53. Part II: Response of tuna to visual and visual-chemical stimuli. U.S. Fish and Wildlife Serv., Spec. Sci. Rept. Fish. 130:63–124.

Hunn, E. S. 1977. Tzeltal folk zoology. Academic Press.

Hurum, H. J. 1977. A history of the fish hook. A. C. Black Ltd., London.

Iles, T. D. 1974. The tactics and strategy of growth in fishes. Pp. 331–346 *in* F. R. Harden-Jones (ed.) Sea fisheries research. Wiley & Sons.

Inanami, Y. 1941. Report of oceanographic changes and fishing conditions in Palau waters. South Sea Fishery News (Nanyō Suisan Jōhō), 5(2):2–6. (In Japanese.)

Inoue, M., R. Amano, Y. Iwasaki, and M. Yamauti. 1968. Studies on environments alluring skipjack and other tunas, VII. On the driftwoods accompanied by skipjack and other tunas. Bull. Jap. Soc. Sci. Fish. 34:288–294.

An investigation of the waters adjacent to Ponape. 1937. Fish. Expt. Stn. Progress Rept. 1: 1923–35. South Seas Government-General, Palau. Translated from Japanese by W. G. Van Camp. Pp. 22–34 *in* U.S. Dept. Interior, Fish Wildlife Serv., Spec. Sci. Rept.: Fisheries, No. 46.

Isaacs, J. D. 1976. Sanity and other factors in aquatic resource development. Pp. 70–85 *in* R. F. Scagel (ed.) Mankind's future in the Pacific. Univ. of British Columbia Press.

Ivens, W. G. 1927. Melanesians of the southeast Solomon Islands. Kegan Paul, Trench, Trubner & Co.

Johannes, R. E. 1975. Pollution and degradation of coral reef communities. Pp. 13–51 *in* E. J. Ferguson Wood and R. E. Johannes (eds.) Tropical marine pollution. Elsevier, Amsterdam.

———. 1977. Traditional law of the sea in Micronesia. Micronesica 13: 121–127.

———. 1978a. Traditional marine conservation methods in Oceania and their demise. Ann. Rev. Ecol. Syst. 9:349–364.

———. 1978b. Reproductive strategies of coastal marine fishes in the tropics. Envir. Biol. Fish. 3:65–84.

———. 1980. Using knowledge of the reproductive behavior of reef and lagoon fishes to improve fishing yields. *In* J. Bardach, J. J. Magnuson, R. May, and J. Reinhart (eds.) Fish behavior and its use in the capture and culture of fishes. I.C.L.A.R.M., Manila.

Johannes, R. E., J. Alberts, C. D'Elia, R. A. Kinzie, III, L. R. Pomeroy, W. Sottile, W. Wiebe, J. A. Marsh, Jr., P. Helfrich, J. Maragos, J. Meyer, S. Smith, D. Crabtree, A. Roth, L. R. McCloskey, S. Betzer, N. Marshall, M. E. Q. Pilson, G. Telek, R. I. Clutter, W. D. DuPaul, K. L. Webb, and

J. M. Wells, Jr. 1972. The metabolism of some coral reef communities: a team study of nutrient and energy flux at Eniwetok. Bio. Sci. 22: 541–543.

Johannes, R. E., and R. Gerber. 1974. Import and export of detritus and zooplankton by an Eniwetok coral reef community. Pp. 97–104 in Proc. 2nd Internat. Symp. Coral Reefs. Volume I.

Kahn, E. J., Jr. 1966. A reporter at large. Micronesia I. New Yorker. June 11: 42–111.

Kalmijn, A. J. 1974. The detection of electrical fields from inanimate and animate sources other than electric organs. Pp. 147–200 in A. Fesard (ed.) Handbook of Sensory Physiology. Vol. 3/3.

———. 1978. Electric and magnetic sensory world of sharks, skates, and rays. Pp. 507–528 in E. S. Hodgson and R. F. Mathewson (eds.) Sensory biology of sharks, skates, and rays. Office of Naval Res., Dept. of the Navy, Arlington, Virginia.

Kanda, T. 1944. Ecological studies on marine algae from Korôru and adjacent islands in the South Sea Islands. Palao Trop. Biol. Stn. Stud. Vol. 4:733–800.

Kaneshiro, S. 1958. Land tenure in the Palau islands. Pp. 288–339 in Land tenure patterns in the trust territory of the Pacific islands. Office of the T.T.P.I. Staff Anthropologist, Guam.

Kawaguchi, K. 1974. Handline and longline fishing explorations for snapper and related species in the Caribbean and adjacent waters. Mar. Fish. Rev. 36:8–31.

Kayser, A. 1936. Die Fischerie auf Nauru (Pleasant Island). Mitt. Anthrop. Gesel. Wien 66:92–131, 149–204.

Kearney, R. E., and J. P. Hallier. 1978. Interim report of the activities of the skipjack survey and assessment program in the waters of New Caledonia. South Pacific Commission Skipjack Survey and Assessment Program Prelim. Country Rept. #3.

Keate, G. 1788. An account of the Pelew Islands situated in the western part of the Pacific Ocean: composed from journals and communications of Captain Henry Wilson and some of the officers, who, in August 1783, were there shipwrecked in the Antelope, a packet belonging to the honorable East India Company. Luke White, Dublin.

Kennedy, D. G. 1929. Field notes on the culture of Vaitupu, Ellice Islands. J. Polyn. Soc. 38(150):Memoir Suppl. 1–38.

Kent, G. 1979. The politics of Pacific island fisheries. Westview Press.

Kishinouye, K. 1923. Contributions to the comparative study of the so-called scombroid fishes. J. Coll. Agric. Tokyo 8:293–475.

Klee, G. A. 1972. The cyclic realities of man and nature in a Palauan village. Ph.D. dissertation, Univ. of Oregon.

Klemmer, K. 1967. Observations on the sea snake *Laticauda laticaudata* in captivity. Intern. Zoo. Yearbook 7:229–231.

Knowles, F. G. W., and T. B. Carlisle. 1956. Endocrine control in crustacea. Biol. Rev. 31:396–3.

Korringa, P. 1947. Relations between moon and periodicity in the breeding of marine animals. Ecol. Monogr. 17:347–381.

Kotzebue, O. von. 1821. A voyage of discovery into the South Seas and the Bering Straits. (Translated from Russian by H. E. Lloyd.) 3 Vols. Longman, Hurst, Rees, Orme and Brown, London.

Kramer, A. 1903. Die Samoan-Inseln. Vol. 2 Part 4. (Translated from German by Anonymous.) Stuttgart.

————. 1929. Palau. Ergebnisse der Siidsee—Expedition 1908–1910, G. Thilenius (ed.) II, B., III, iii, 1–362. (Translated from German by Anonymous). Friedrichsen & Co. Hamburg.

Kubary, J. S. 1895. Ethnographishe Beitrage zur Kenntnis des Karolinen Archipels. (Translated from German by Anonymous.) Trap, Leiden.

Laevastu, T. 1962. The causes and predictions of surface current in sea and lake. Hawaii Instit. Geophys. Rept. 21. 55p.

LeBar, F. M. 1952. The material culture of Truk. Yale Univ. Publ. Anthrop. No. 68.

Lessa, W. A. 1975. The Portuguese discovery of the Isles of Sequeria. Micronesica 11:35–70.

Lewis, R. 1972. We, the navigators. Univ. Press of Hawaii.

Lieber, A. L. 1978. The lunar effect. Anchor Press.

Limbaugh, C. 1963. Field notes on sharks. Pp. 63–94 *in* P. W. Gilbert (ed.) Sharks and survival. D. C. Heath Co.

Lindsey, C. C. 1972. Fisheries training in the region served by the University of the South Pacific. University of the South Pacific, Suva, Fiji.

Longuet-Higgins, M. S., M. E. Stern, and H. Stommel. 1954. The electrical field induced by ocean currents and waves with application to the method of towed electrodes. Pap. Phys. Oceanogr. 13:1–37.

McClintock, M. K. 1971. Menstrual synchrony and suppression. Nature. 229:244–245.

McCoy, M. A. 1974. Man and turtle in the central Carolines. Micronesica 10:207–221.

McDonald, R. L. 1966. Lunar and seasonal variations in obstetric factors. J. Genetic. Psychol. 108:81–87.

MacGregor, G. 1937. Ethnology of Tokelau Islands. B. P. Bishop Mus. Bull. 146.

McKnight, R. K. 1972. Rigid models and ridiculous boundaries. Political development and practice in Palau—circa 1955–64. Micronesica 8:23–35.

Major, H. 1939. Salt water fishing tackle. Funk & Wagnalls.

Malinowski, B. 1918. Fishing and fishing magic in the Trobriand Islands. Man 18(53):87–92.

Manacop, P. R. 1937. The artificial fertilization of dangit, *Amphacanthus oramin* (Bloch and Schneider). Philippine J. Sci. 62:229–237.

Matsumoto, W. M. 1952. Experimental surface gill net fishing for skipjack (*Katsuwonus pelamis*) in Hawaiian waters. U.S. Dept. Interior Fish Wildlife Spec. Sci. Rept: Fisheries. 90.

May, R. C., G. S. Akiyama, and M. T. Santerre. 1979. Lunar spawning of the threadfin *Polydactylus sexfilis*, in Hawaii. Fish. 76:900–904.

Menaker, W., and A. Menaker. 1959. Lunar periodicity in human reproduction: a likely unit of biological time. Am. J. Obstet. Gynec. 77:905–914.

Military geology of Palau Islands, Caroline Islands. 1956. U.S. Geological Surv.

Miller, J. M. 1973. Nearshore distribution of Hawaiian marine fish larvae: effects of water quality, turbidity and currents. Pp. 217–231 *in* J. H. S. Blaxter (ed.) The early life history of fish. Springer-Verlag, Berlin.

Minei, H. 1966. Studies on the land crabs (family Geocarcinidae) in the Ryukyu Islands. Biol. Mag. Okinawa 3:8–10. (in Japanese)

Morrill, W. T. 1967. Ethnoichthyology of the Cha-Cha. Ethnology 6:405–416.

Moss, S. A., III. 1965. The feeding mechanisms of three sharks: *Galeocerdo cuvieri* (Peron & Le Sueur), *Negaprion brevirostris* (Poey), and *Ginglymostoma cirratum* (Bonnaterre). Ph.D. dissertation, Cornell Univ.

Moulton, J. M. 1958. The acoustical behavior of some fishes in the Bimini area. Biol. Bull. 114:357–374.

Munro, J. L. 1974a. The biology, ecology, exploitation and management of Caribbean reef fishes. Part. Vm. Summary of biological and ecological data pertaining to Caribbean reef fishes. Res. Rept. #3 Zoology Dept., Univ. West Indies.

———. 1974b. The biology, ecology, exploitation and management of Caribbean reef fishes. Part 1. Coral reef fish and fisheries of the Caribbean Sea. Res. Rept. #3 Zoology Dept., Univ. West Indies.

Munro, J. L., P. H. Reeson, and V. G. Gaut. 1971. Dynamic factors affecting the performance of the Antillean fish trap. Proc. Gulf Carib. Fish. Instit. 23:184–194.

Munz, F. W., and W. N. McFarland. 1973. The significance of the spectral position in the rhodopsins of tropical marine fishes. Vision Res. 13:1829–1874.

———. 1975. Presumptive cone pigments extracted from tropical marine fishes. Vision Res. 15:1045–1062.

Murai, M., F. Pen, and C. D. Miller. 1958. Some tropical South Pacific foods. Univ. of Hawaii Press.

Murphy, G. I., and I. I. Ikehara. 1955. A summary of sightings of fish schools and bird flocks and of trolling in the central Pacific. Spec. Sci. Rept. U.S. Fish. Wildlife Serv., Fish. 154.

Myrberg, A. A., Jr. 1978. Underwater sound—its effects on the behavior of sharks. Pp. 391–417 *in* E. S. Hodgson and R. F. Mathewson (eds.) Sensory biology of sharks, skates and rays. Office of Naval Research, Dept. of the Navy, Arlington, Virginia.

Myrberg, A. A., Jr., C. R. Gordon, and A. P. Klimley. 1976. Attraction of free ranging sharks by low frequency sound. Pp. 205–228 *in* A. Schuijf and A. D. Hawkins (eds.). Sound reception in fish. Elsevier, Amsterdam.

Myrberg, A. A., Jr., S. J. Ha, S. Walewski, and J. C. Bandberry. 1972. Effectiveness of acoustic signals in attracting epipelagic sharks. Bull. Mar. Sci. 22:926–949.

Nakayama, M., and F. L. Ramp. 1974. Micronesian navigation, island empires and traditional concepts of ownership of the sea. 5th Congress Micronesia, Saipan, Northern Marianas Islands.

Nash, C. E., and Kuo, Ching-Ming. 1976. Preliminary capture, husbandry and induced breeding results with the milkfish, *Chanos chanos* (Forskål). Pp. 139–159 *in* Proc. Internat. Milkfish Conf., SEAFDEC.

Neill, S. R. St. J. 1966. Observations on the behavior of the grouper species *Epinephelus guaza* and *E. alexandricus (Serranidae)*. Pp. 101–106 *in* J. N. Lythgoe and J. D. Woods (eds.) Underwater Association Rept. 1966–67. T. G. W. Industrial and Research Promotions Ltd.

Nelson, D. R., and R. H. Johnson. 1976. Some recent observations on the acoustic attraction of Pacific reef sharks. Pp. 229–237 *in* A. Schuijf and A. D. Hawkins (eds.) Sound reception in fish. Elsevier, Amsterdam.

Nelson, R. K. 1969. Hunters of the northern ice. Univ. Chicago Press.

Nielsen, L. A. 1976. The evolution of fisheries management philosophy. Mar. Fish. Rev. 38:15–23.

Nietschmann, B. 1973. Between land and water. Seminar Press.

Nilsson, M. P. 1920. Primitive time-reckoning. Oxford Univ. Press.

Nomura, M. 1980. Selected traits of fish behavior and the design of fishing gear. *In* J. Bardach, J. Magnuson, R. May, and J. Reinhart (eds.) Fish behavior and fisheries management (capture and culture). I.C.L.A.R.M, Manila.

Nordhoff, C. B. 1928. Fishing for the oilfish. Native methods of deepsea fishing for *Ruvettus pretiosus* at Atiu, Hervey Group and elsewhere in the South Seas. Nat. Hist. 28:40–45.

———. 1930. Notes on the offshore fishing of the Society Islands. J. Polynesian Soc. 39:137–173, 221–262.

Norse, E. A., and V. Fox-Norse. 1977. Studies on portunid crabs from the eastern Pacific. II. Significance of the unusual distribution of *Euphylax dovii*. Mar. Biol. 40:374–376.

O'Connell, J. F. 1836. A residence of eleven years in New Holland and the Caroline Islands. B. B. Mussey, Boston.

Oliver, D. 1974. Ancient Tahitian society. Vol. 1. Univ. of Hawaii Press.

Ommaney, F. D. 1966. A draught of fishes. Crowell Co.

Ong, K. H. 1966. Observations on the post-larval life history of *Scylla serrata* Forskål, reared in the laboratory. Malays. Agric. J. 45:429–443.

Orbach, M. K. 1977. Hunters, seamen and entrepreneurs. Univ. of California Press.

Osborne, D. 1966. The archaeology of the Palau Islands. B. P. Bishop Mus. Bull. 230.

Osley, M., D. Summerville, and L. B. Borst. 1973. Natality and the moon. Am. J. Obstet. Gynec. 117:413–15.

Ottino, P., and Y. Plessis. 1972. Les classifications ouest Paumotu de quelques poissons scaridés et labridés. Pp. 361–371 *in* J. M. C. Thomas and L. Bernot (eds.) Lanques et Techniques Nature et Société. Vol. II. Editions Klinksieck, Paris.

Owings, R. D., and R. G. Coss. 1977. Snake mobbing by California ground squirrels: adaptive variation and ontogeny. Behaviour 62:50–69.

Pfeiffer, W. 1963. Alarm substances. Experientia 19:113–123.

Popper, D., R. C. May, and T. Lichatowich. 1976. An experiment in rearing larval *Siganus vermiculatus* (Valenciennes) and some observations on its spawning cycle. Aquaculture 7:281–290.

Popper, D., R. Pitt, and Y. Zohar. 1979. Experiments on the propagation of Red Sea siganids and some notes on their reproduction in nature. Aquaculture 16:177–182.

Porter, J. W., and K. Porter. 1973. The effects of Panama's Cuna Indians on coral reefs. Discovery 8:1363–1383.

Pratt, A. E. 1906. Two years among the New Guinea cannibals. London.

Pritchard, P. C. H. 1977. Marine turtles of Micronesia. Chelonia Press.

Pritchard, W. T. 1866. Polynesian reminiscences; or life in the South Pacific islands. London.

Randall, J. E. 1961a. Observations on the spawning of surgeon fishes (Acanthuridae) in the Society Islands. Copeia 1961: 237–238.

———. 1961b. A contribution to the biology of the convict surgeonfish of the Hawaiian Islands, *Acanthurus triostegus sandvicensis*. Pac. Sci. 15:215–272.

———. 1966. The living javelin. Sea Frontiers 6:228–233.

———. 1973. Tahitian fish names and a preliminary checklist of the fishes of the Society Islands. Occas. Pap. B. P. Bishop Mus. 27:167–214.

Randall, J. E., and V. E. Brock. 1960. Observations on the ecology of epinepheline and lutjanid fishes of the Society Islands, with emphasis on food habits. Trans. Amer. Fish. Soc. 89:9–16.

Randall, J. E., and G. S. Helfman. 1973. Attacks on humans by the blacktip reef shark (*Carcharhinus melanopterus*). Pac. Sci. 27:226–238.

Randall, J. E., and H. A. Randall. 1963. The spawning and early development of the Atlantic parrotfish, *Sparisoma rubripinna*, with notes on other scarid and labrid fishes. Zoologica 48:49–60.

Randall, R. A. 1977. Change and variation in Samal fishing: making plans to "make a living" in the Southern Philippines. Ph.D. dissertation, Univ. of California, Riverside.

Reaka, M. L. 1976. Lunar and tidal periodicity of molting and reproduction in stomatopod crustacea: a selfish herd hypothesis. Biol. Bull. 150:468–490.

Reichel-Dolmatoff, G. 1974. Amazonian cosmos. Univ. of Chicago Press.

Reighard, J. 1920. The breeding behavior of the suckers and minnows. I. The suckers. Biol. Bull. 38:1–32.

Reinman, F. M. 1967. Fishing: an aspect of oceanic economy. Fieldiana: Anthropology 56:94–208.

Report of survey of fishing grounds and channels in Palau waters 1925–26. 1937. Fish. Res. Stn. Prog. Rept. No. I, 1923–1935. South Seas Government-General, Palau. Translated from Japanese by W. G. Van Camp. Pp. 11–22 *in* U.S. Dept. Interior, Fish Wildl. Serv., Spec. Sci. Rept. Fisheries, No. 42.

Reshetnikov, Yu. S., and R. M. Claro. 1976. Cycles of biological processes in tropical fishes with reference to *Lutjanus synagris*. J. Ichthyol. 16:711–721.

Robertson, D. R., and S. G. Hoffman. 1977. The roles of female mate choice and predation in the mating systems of some tropical labroid fishes. Z. Tierpsychol. 45: 298–320.

Roede, M. J. 1972. Color as related to size, sex and behavior in seven Caribbean labrid fish species. Stud. Fauna Curaçao 42:1–264.

Romilly, H. H. 1856. The western Pacific and New Guinea. John Murray, London.

Roughley, T. C. 1966. Fish and fisheries of Australia. Angus & Robertson, Sydney.

Royce, W. F., and T. Otsu. 1955. Observations on skipjack schools in Hawaiian waters. Spec. Sci. Rept. U.S. Fish and Wildlife Serv. Fish. 147.

Saetersdal, G. 1963. Selectivity of long-liners. I.C.N.A.F. Spec. Publ. #5. Pp. 189–192.

Safford, W. 1905. The useful plants of Guam. Contrib. U.S. Natn'l Herbarium 9.

Sahlins, M. D. 1974. Stone age economics. London.

Saigusa, M., and T. Hidaka. 1978. Semilunar rhythm in the zoea-release activity of the land crabs *Sesarma.* Oecologia 37:163–176.

Sailing Directions for the Pacific Islands. 1964. Volume I. Western groups, including the Solomon Islands. U.S. Naval Oceanographic Office, Government Printing Office.

Sale, P. F. 1970. Distribution of larval Acanthuridae off Hawaii. Copeia 1970:765–766.

———. 1978. Coexistence of reef fishes—a lottery for living space. Env. Biol. Fish. 3:85–102.

Sarfert, E. 1919. Kusaie. Ergebnisse der Südsee—Expedition 1908–1910, ed. 6. Thilenius, II, B, IV 645 p. Friedrichsen, Hamburg.

Schöne, H. 1963. Menotaktische Orientierung nach polarisiertem und unpolarisiertem Licht bei der Mangrovekrabbe *Goniopsis.* Z. Vergl. Physiol. 46:496–514.

Schott, G. 1939. Die äquatorialen Stromungen des Westlichen Stillen Ozeans. Ann. d. Hydrogr. u. Mar. Meteorol. 68:247–257.

Seidel, H. 1905. Die Bewohner der Tobi—Insel. Globus. 87:113–117.

Semper, K. 1873. Die Palau—Inseln im Stillen Ozean. Leipzig.

Seurat, L. G. 1905. Les engines de pêche des anciens Paumotu. L'Anthropologie 16:295–307.

Severance, C. J. 1976. Land, food and fish: strategy and transaction on a Micronesian atoll. Ph.D. dissertation, Univ. of Oregon.

Shokita, S. 1971. On the spawning habits of the land crab. *Cardisoma hirtipes* Dana from Ishigaki Island in the Ryukyu Islands. Biol. Mag. Okinawa 7:27–32. (in Japanese; English summary.)

Shul'man, G. E. 1974. Life cycles of fish. Keter Publishing House, Jerusalem. (English translation distributed by Halstead Press.)

Silas, E. G. 1963. Synopsis of biological data on double-lined mackerel *Grammatorcynus bicarinatus* (Quoy and Gaimard) (Indo-Pacific). F.A.O. Fish. Rept. 6:811–833.

Smith, C. L. 1972. A spawning aggregation of Nassau grouper *Epinephelus striatus* (Bloch). Trans. Amer. Fish. Soc. 101:257–261.

Smith, R. O. 1947. Survey of the fisheries of the former Japanese mandated islands. U.S. Dept. Interior Fish and Wildl. Serv. Fishery Leaflet 273.

Smith, S. V. 1978. Coral-reef area and the contributions of reefs to processes and resources of the world's oceans. Nature 273(5659):225–226.

Springer, S. 1960. Natural history of the sandbar shark, *Eulamia milberti*. U.S. Fish and Wildlife Serv., Fish. Bull. 178. Vol. 61:1–38.

Starck, W. A., II, and R. E. Schroeder. 1970. Investigations on the grey snapper, *Lutjanus griseus*. Stud. Trop. Oceanogr. #10.

Stevenson, D. K. 1977. Management of a tropical pot fishery for maximum sustainable yield. Proc. 30th Ann. Mtg. Gulf and Carib. Fish. Instit. pp. 95–115.

Stevenson, D. K. and N. Marshall. 1974. Generalisations on the fisheries potential of coral reefs and adjacent shallow water environments. Pp. 147–156. Proc. 2nd Internat. Symp. Coral Reefs. Vol. I.

Stevenson, R. A., Jr., 1972. Regulation of feeding behavior of the bicolor damselfish (*Eupomacentrus partitus* Poey) by environmental factors. Pp. 278–302 *in*: H. E. Winn and B. L. Olla (eds.) Behavior of marine animals. Vol. 2 vertebrates. Plenum.

Stimson, J. F. 1928. Tahitian names for the nights of the moon. J. Polyn. Soc. 37:326–337.

Stone, B. C. 1970. The flora of Guam. Micronesica 6:1–659.

Strasburg, D. W., and H. S. H. Yuen. 1958. Preliminary results of underwater observations of tuna schools and practical applications of these results. Proc. Indo-Pacific Fish. Council 8(3):84–89.

Sutcliffe, W. H., Jr. 1956. Effect of light intensity on the activity of the Bermuda spiny lobster *Panulirus argus*. Ecol. 37:200–201.

Talbot, F. H., and F. Williams. 1956. Sexual color differences in *Caranx ignobilis* (Forsk). Nature 178:934.

Tampi, P. R. S. 1957. Some observations on the reproduction of milkfish *Chanos chanos* (Forskål). Proc. Ind. Acad. Sci. 46:254–273.

Tanaka, S. K. 1973. Suction feeding by the nurse shark. Copeia 1973: 606–608.

Taylor, C. B. 1957. Hawaiian Almanac. Tongg Publ. Co. Ltd.

Thomas, A. J. 1972. Crab fishery of the Pulicat Lake. J. Mar. Biol. Assoc. India 13:278–280.

Thompson, J. M. 1955. The movements and migrations of mullet (*Mugil cephalus* L.). Austr. J. Mar. Freshw. Res. 6:328–347.

Thompson, R., and J. L. Munro. 1974. The biology, ecology and bionomics of Caribbean reef fishes: Serranidae (hinds and groupers). Res. Rept #3 Zoology Dept., University of the West Indies. Part Vb.

Titcomb, M. 1972. Native use of fish in Hawaii. Univ. of Hawaii Press.

Tjiptaminoto, R. M. 1956. Some notes on adult *Chanos*. Proc. Indo. Pac. Fish. Counc., 5th Ses. 2–3:209–210.

Tolerton, B., and J. Rauch. 1949. Social organization, land tenure and subsistence economy of Lukunor, Nomoi Islands. Pacific Science Board, Washington. CIMA Rept. No. 26.

Tranter, D. J., and J. George. 1972. Zooplankton abundance at Kavarati and Kalpeni Atolls in the Laccadives. Pp. 239–256 in Proc. Symp. Corals and Coral Reefs 1969, Mar. Biol. Assoc. India.

Tsuda, R. T., and P. G. Bryan. 1973. Food preference of juvenile *Siganus rostratus* and *S. spinus* in Guam. Copeia 1973:604–606.

Turbott, I. G. 1950. Fishing for flying fish in the Gilbert and Ellice Islands. J. Polynes. Soc. 59:349–367.

Useem, J. 1945. The changing structure of a Micronesian society. Amer. Anthrop. 47:567–588.

———. 1949. Report on Palau. Pacific Science Board, Washington, D.C., CIMA Rept. No. 21.

Useem, J. 1955. Palau. Pp. 126–150 in M. Mead (ed.) Cultural patterns and technical change. New American Library.

Vaea, Hon., and W. Straatmans. 1954. Preliminary report on fisheries survey in Tonga. J. Polyn. Soc. 63:199–215.

Van Meir, L. W. 1973. Problems in implementing new fishery management programs. A. A. Sokoloski (ed.) Pp. 9–11 in Ocean fishery management. NOAA Techn. Rept. NMFS Circ. 371.

Veerannan, K. M. 1974. Respiratory metabolism of crabs from marine and estuarine habitats: an interspecific comparison. Mar. Biol. 26:35–43.

Villadolid, D. U. 1940. Philippine fisheries and problems of their conservation. Proc. 6th Pac. Sci. Congr. 3:369–385.

Vivien, M. L. 1973. Contribution a la connaissance de l'éthologie alimentaire de l'ichthyofaune du platier interne des récife coralliens de Tuléar (Madagascar). Tethys 5(suppl.):221–308.

Wagner, D. P., and R. S. Wolf. 1974. Results of troll fishing explorations in the Caribbean. Mar. Fish. Rev. 36(9):35–43.

Warner, R. R., and D. R. Robertson. 1978. Sexual patterns in the labroid fishes of the western Caribbean, I: The wrasses (Labridae). Smithsonian Contrib. Zool. 254:1–27.

Wass, R. C. 1973. Size, growth and reproduction of the sandbar shark, *Carcharhinus milberti* in Hawaii. Pac. Sci. 27:305–318.

Waterman, T. H., and K. W. Horch. 1966. Mechanism of polarized light perception. Science 154:467–475.

Westenberg, J. 1953. Acoustical aspects of some Indonesian fisheries. J. du Cons. 18:311–325.

Westernhagen, H. von. 1974. Observations on the natural spawning of *Alectis indicus* (Rüppell) and *Caranx ignobilis* (Forsk.) (Carangidae). J. Fish. Biol. 6:513–516.

Whitmee, S. J. 1875. On the habits of *Palolo viridis* – Proc. Zool. Soc. (London) 1875:496–502.

Williams, F. 1965. Further notes on the biology of the East African pelagic fishes of the families Carangidae and Sphyraenidae. E. Afric. Agric. Forestry J. 31:141–168.

Williams, G. C. 1959. Ovary weights of darters: a test of alleged association of parental care with reduced fecundity in fish. Copeia 1959:18–24.

Wilson, J. 1759. A missionary voyage to the southern Pacific Ocean. London.

Woodworth, W. M. 1903. Preliminary report on the "Palolo" worm of Samoa, *Eunice viridis* (Gray). Amer. Nat. 37:875–881.

———. 1907. The palolo worm, *Eunice viridis* (Gray). Bull. Mus. Comp. Zool., Harvard Coll. 51:3–21.

Wyrtki, K. 1974. Sea level and seasonal fluctuations of the equatorial currents in the western Pacific Ocean. J. Phys. Oceanogr. 4:91–103.

Yamanouchi, T. 1955. On the poisonous substance contained in holothurians. Publ. Seto Mar. Biol. Lab. 4:183–203.

Yanaihara, T. 1940. Pacific islands under Japanese mandate. Oxford Univ. Press.

Yuen, H. S. H. 1970. Behavior of skipjack tuna, *Katsuwomus pelamis*, as determined by tracking with ultrasonic devices. J. Fish. Res. Bd. Can. 27:2071–2079.

Znamierowska-Prüffer, M. 1966. Thrusting implements for fishing in Poland and neighboring countries. (Translated from Polish.) Sci. Publications Co-operation Center, Central Instit. Sci. Tech. Econ. Inform., Warsaw, Poland.

Zucker, N. 1978. Monthly reproductive cycles in three sympatric hood-building tropical fiddler crabs (genus *Uca*). Biol. Bull. 155:410–424.

INDEX

Tabar Island, viii
Taboos, fishing, 14, 88
Tag and recapture studies,
   57, 82
Tahiti(ans), viii, ix, 24,
   32−35, 39, 60, 62, 111,
   113−115, 156, 162, 164,
   166, 168, 173, 179, 185,
   193
Taiwan, 88, 155
Talitrus saltator, 39
Tang, sailfin, 129
Tanganangan, 128
Taod, 11, 17
Tarawa, 147, 152, 157, 158,
   160, 167, 176, 185
Tariff, protective, 80
Taro, 5
Tattoo: on Palauan shark
   fishermen, 14; Pulo
   Annan, 130; Tobian, 130
Tauchacha, 129
Tavitef, 129
Taxonomy, fish, 124−130;
   folk, 120−130
Tebukbuk, 27
Technology, imported, 148
Teeth: of needlefish, 93;
   pharyngeal, of jacks, 19; of
   triggerfish, 117
Teke titi, 58
Tekoi 'l chei, 3
Temaubour, 124
Temol, 184
Tenure, reef and lagoon,
   64−66, 70, 77−79
Terekrik, 140, 187, 188
Teribour, 129
Terminalia catappa, 14
Tern: black noddy, 60−62,
   107; brown, 107; white,
   61, 62, 107
Ternate, 100, 119
Teter, 127
*Thalassosteus appendi-
   culatus*, time and location
   of spawning of, 160
Thermocline: effect of depth

of on purse seining for
   tuna, 97; effect of on reef
   fish movements, 21
Thunnus albacares, 61
Tides, 29, 55; currents
   associated with, 36, 52;
   ebbing, 15−17, 35, 36, 49,
   55, 153, 186; effects of on
   fishing of, 32, 49−58;
   height of, 32, 50; neap, 35,
   50, 52, 53, 55, 193; relation
   of to moon phase, 32, 50;
   rising, 49, 50, 52, 53, 55,
   186; slack, 50, 52; spring,
   17, 35, 39, 50, 55, 57, 192,
   193; and time of birth, 35;
   and time of death, 35;
   timing of, 32, 50
Tkuu, 61
Tmur ra ongos, 43
Toachel m'lengui, 180, 186,
   187
Tobi, 39, 43, pl. 16; area of,
   86; fishing methods of,
   90−100; fish names of,
   125−130; food supply on,
   86; language of, 96;
   learning to fish on, 88;
   location of, 86; population
   of, 86; potential fish yield
   from reefs of, 87; religion
   of, 88; reputation of
   natives of for ferocity, 96
Tokelaus, 58
Tonga, 192
Tool, use of by triggerfish,
   189
Torch fishing, 90, 92, 99
Tourism in Pacific islands,
   148
Toxin, nerve, from sea
   cucumber, 12
Toxotes jaculatrix, 141
Trachurus boops, 140; best
   fishing times for, 188;
   fishing pressure on, 188;
   food of, 187; time and
   location of spawning of, 187